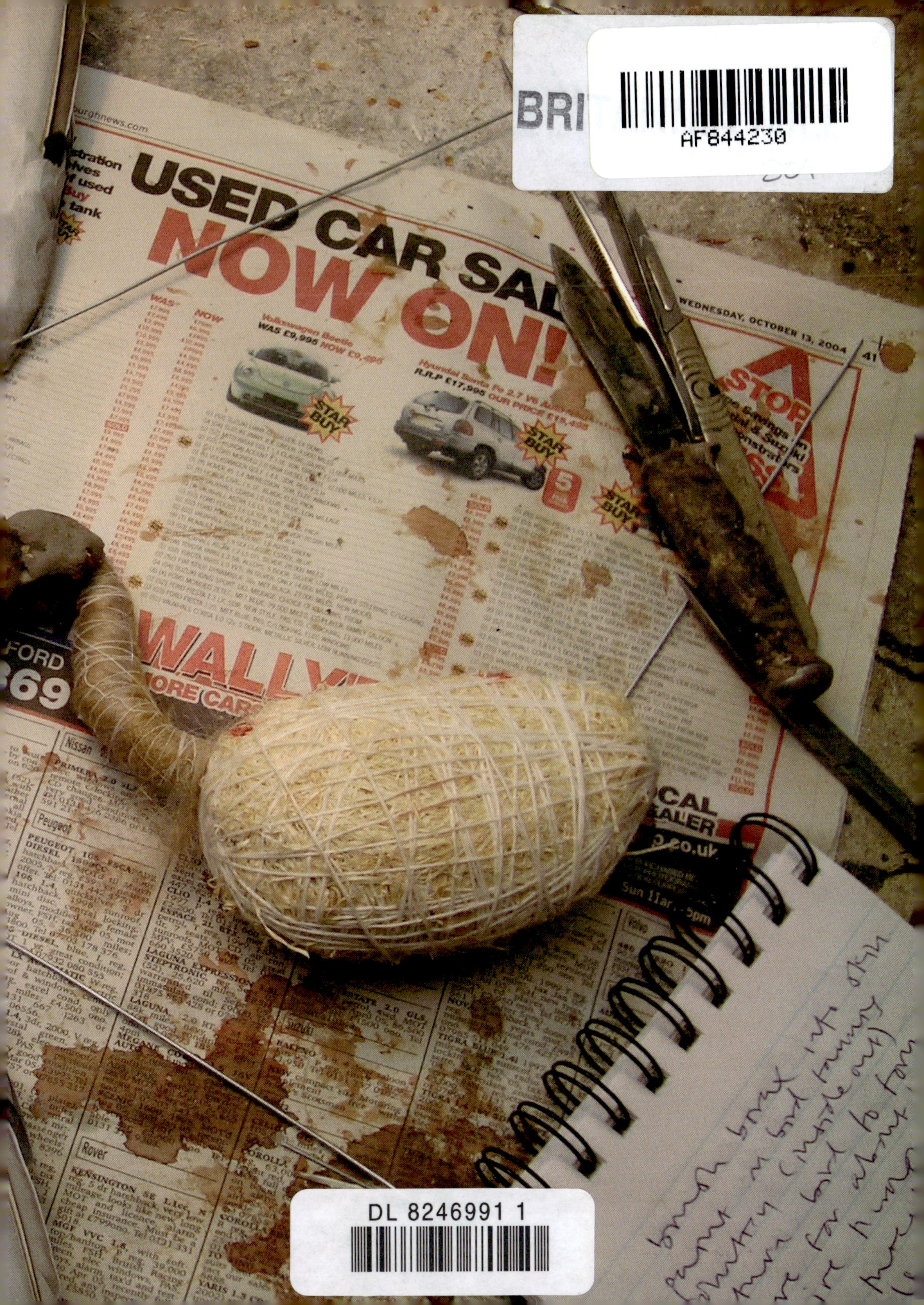
USED CAR SA
NOW ON!
WEDNESDAY, OCTOBER 13, 2004
STAR BUY
WALLY
FORD
Nissan
Peugeot
Rover

Polly Morgan

Psychopomps

Haunch of Venison

Psychopomp

= =

– *noun* a conductor of spirits or souls to the other world, as Hermes or Charon.

- -

For my Mother and Father

- -

Polly Morgan is a member of the guild of taxidermists, which adheres to and promotes the law pertaining to taxidermy, and keeps a detailed log of all animals in stock.

All taxidermied animals are either road casualties or have been donated to the artist by pet owners and vets after natural or unpreventable deaths.

www.pollymorgan.co.uk

Contents

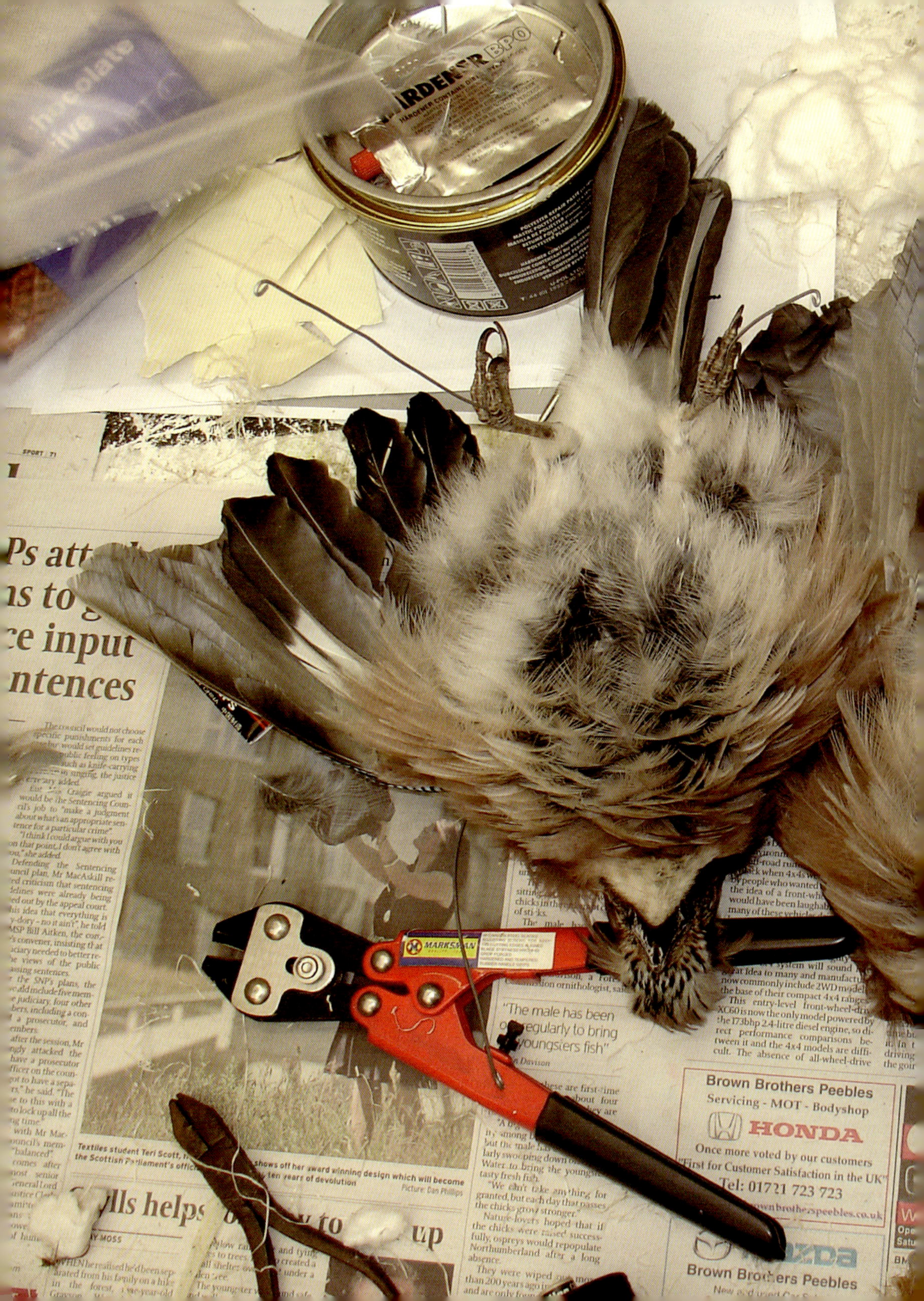
HARDENER BPO
Ps att
ce input
ntences
"The male has been
regularly to bring
youngsters fish"
MARKSMAN
Brown Brothers Peebles
Servicing - MOT - Bodyshop
HONDA
Once more voted by our customers
"First for Customer Satisfaction in the UK"
Tel: 01721 723 723
Brown Brothers Peebles

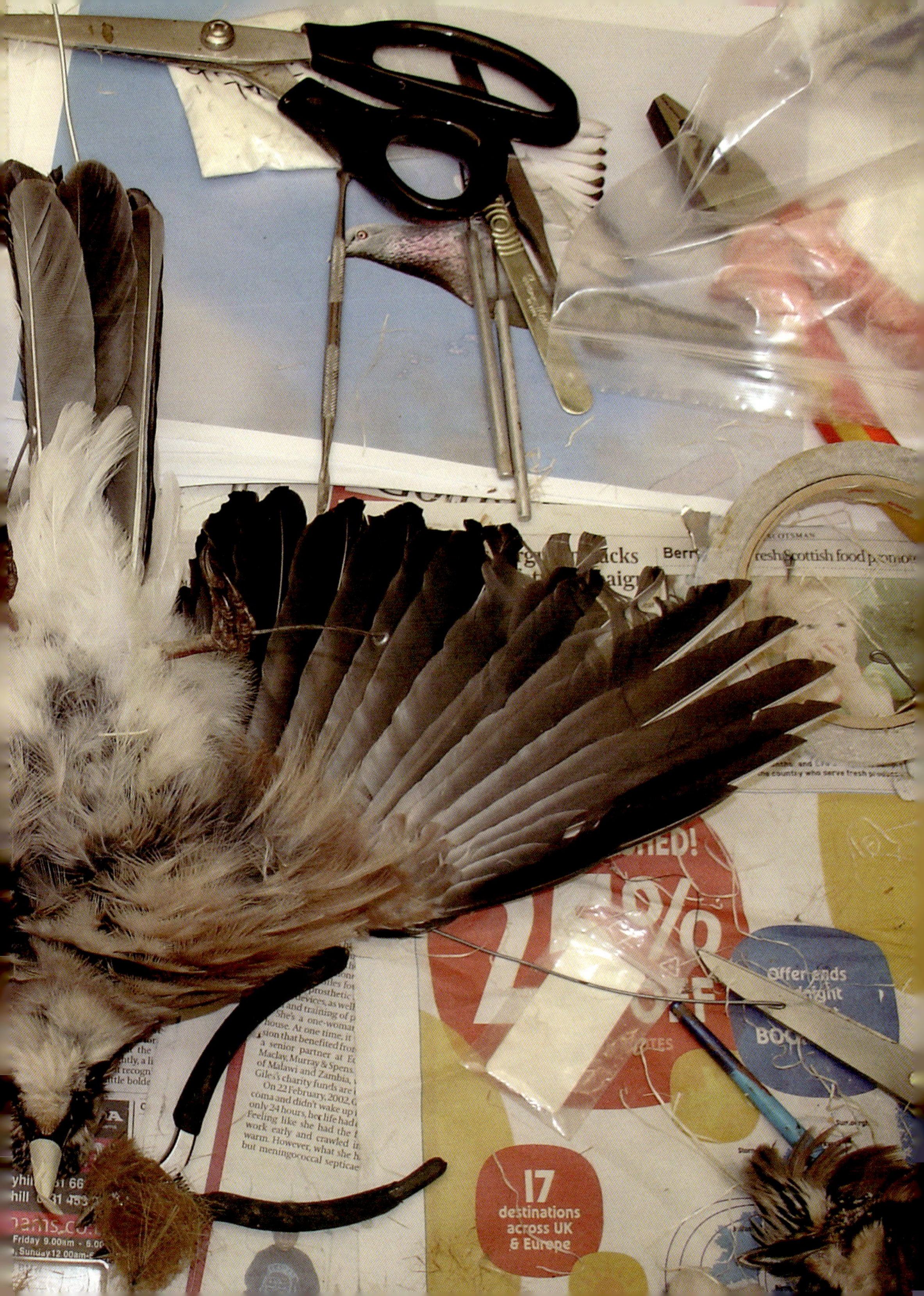
17
destinations
across UK
& Europe

Introduction: Death as a Psychopomp

Tom Hunt

The title of this book comes from Polly Morgan's latest series of works, a group of four hanging mobile sculptures, each of which focuses on the taxidermied bodies of dead birds set within various playful and macabre constructions.

Psychopomps is a Greek word that refers to the guides who lead souls from death to the after-life. In mythology, they include the wing-footed messenger God Hermes and Charon who guards the gates of the Underworld. In Morgan's works, they are played by canaries, finches, crows, pigeons and cardinals, themselves already dead, but suspended as if mid-flight.

The move towards airborne sculpture and Greek mythology, away from glass domed Victoriana, represents a significant shift in attitude as well as aesthetic. The ornamental forms and gothic iconography of Morgan's earlier plinth-based work were founded on upturning prevailing attitudes towards a fusty and even morally dubious medium. The uncanny, doll's house charm of rats in champagne glasses and robins lying on miniature books has been replaced by something more physically and conceptually expansive, by monumental structures and a human scale.

Aside from their role as psychopomps, these works are linked by their titles, each of which reference and puns on diseases common to humans and birds: *Atrial Flutter*, *Systemic Inflammation*, *Blue Fever* and *Black Fever*. This morbid taxonomy introduces an interesting, unremarked aspect of Morgan's work: the near total absence of the human form. We look on limbs, eyes, mouths, limp, dead bodies, sometimes we perceive cruelty or tenderness, vulnerability and exposure, but in each case the part and the effect belongs to a creature unlike us. For an artist so evidently fascinated by death – which is a humanistic idea as well as a technical term – our exclusion is strange, almost phobic, suggesting perhaps a desire to locate oneself outside one's own body.

In *Atrial Flutter* Morgan introduces a human body part for the first time. It is a ribcage, mere carrion, and has been cast rather than stitched together from death. Inside this holey white chamber and placed where the heart might have been, is a small red cardinal, affecting to be a palpitating heart.

Systemic Inflammation shows a burnt flying machine borne in the air by a troupe of finches and canaries. It was inspired by an anonymous illustration published in The Scientific American in 1865 for a fantasy invention, showing a man locked inside a cage being carried aloft by birds. In Morgan's version, the cage remains but there is no man. Why has she chosen to exclude this central figure? Aside from the potentially precarious ethics of taxidermy humans, the man is removed because to include it would compromise the tension between these works as metaphorical *and* factual representations of death.

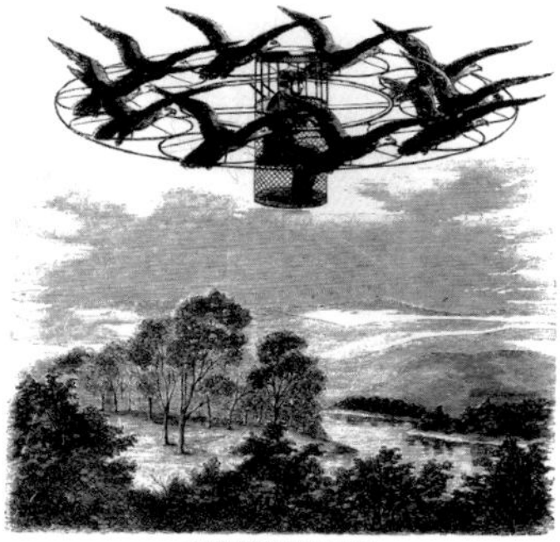

A natural flying-machine [*1865*]

Though death is a pre-requisite for Morgan's taxidermy work, it is offered both literally through the corpse which really is there and dead, and symbolically, via the apparatus that elevates, even ennobles, the dead body. In *Systemic Inflammation,* flight as an analogy for passage and metamorphosis, is crystallised by the hovering birds which carry off the cage. That these birds are dumb to their new-found mythical status does not reduce the effect of the work, but rather signifies the way a sense of the infinite – the sky, flight, immortality – can be inspirational.

The abundance of flapping wings adds a surreal, nightmarish dimension to *Black Fever* and *Blue Fever.* Covered in the wings of a crow or pigeon, planted one on top of the other, each successive wing unfurls to display – as if this were a flick book or an Eadweard Muybridge photo series – the full range of the bird's motion in flight.

While the dynamism of the winged sculptures and the flying machine recall Muybridge, and the fabulous grotesquerie places Morgan's work alongside fellow taxidermist Thomas Grünfeld, there is a mysterious theatre, an affinity with Greek myth and an oblique intimacy which is redolent of the illustrations of Jean Cocteau. She shares with him an ability to render life and death aphoristically, in a condensed form of visual poetry.

Using feathers, skin, leather, resin, wire and steel, Morgan has created a body of work which positions the cadaver above our heads, spectrally and physically, so engaging the history of the hanging mobile via a sustained meditation on death. Like a blue-black night these works are dark and very beautiful.

Polly Morgan's Peaceable Kingdom

= =

Anthony Haden-Guest

Polly Morgan sorts through a consignment of new material, a bag of small dead birds from a pet shop in Enfield. She picks out four and puts two down on a double page of the Sunday Times spread out on the breakfast table. "I'm going to dye them orange," she says, "I'm using womens' hair dye."

She sets to work on a canary, fingers working busily with a scalpel. She has calipers and a spool of black thread beside her and has laid out cotton buds and dabs of auto-body filler on a DIY palette.

Finally she extracts the skinned corpse. It is the shape of a large prune and the color of a three day bruise.

Doesn't it smell?

"It's pretty fresh, this one. Sometimes they do."

She picks and prods and pokes.

"I'm going to skin four today. Then I'm going to bind their bodies with that stuff, woodwool."

These are the wiry shavings used, for instance, to pack china.

Pick, prod, poke.

"I'm stripping the bones of flesh now. If you don't take all the flesh off it will dry out eventually but it could attract flies during that time. Next I'll remove the fat; if I don't do this, the skin could rot and the feathers would fall out."

She points the scalpel at some tiny mounds, melting into pools of liquid jelly. "These are the eyeballs... and the brains... that's the gooey stuff..."

The goo is pinkish-white. It has precisely the soft runniness of the flesh of a sea urchin.

Watching an artist at work is usually interesting but it is not often quite as interesting as this. After a second bird the morning's work was done. Morgan wrapped her materials in plastic and popped them into the freezer alongside a container of Gino's pepperoni pizza, a clump of Helen Browning's organic sausages and three tubs of Häagen-Dazs.

- - - -

Artists tend to be guarded when discussing what originally drove them towards the use of particular forms, content or materials; understandably so because the truest of stories begin to sound hollow to the teller when too often repeated. Morgan is sensitive to this phenomenon of reality-leakage and rejects the notion that there is a golden key which unlocks the door of meaning. So what follows is just part of the story. And that all that matters in the end is the work.

"I grew up in the Cotswolds about twenty miles from Oxford" she says. "We were always surrounded by animals: 200 goats, llamas. dogs, cats, chickens, fish, hamsters. We didn't have much money but had a nice house and ran around after the animals all day."

Occasionally one of these animals would come to a natural or a sticky end and when she was about three, Morgan began collecting the tiniest of the furry or feathered fallen. "My mother explained that if I didn't bury them they would rot".

So the little burials began.

"I used to use matchboxes or cardboard boxes or whatever I could find to put them in. I remember going shopping and my sister had a headache. My Mum got a box of Anadin from her bag and opened it to slide the pills out. Instead of pills a hard, dead hamster slipped out onto her palm. 'Oh dear. Wrong box,' she said."

Animals continued to inhabit Morgan's imagination through her adolescence. "I'm drawn to dusty antique shops and Cabinets of Curiosities." It was in this way that she became familiar with the vivid afterlife granted to animals by the taxidermist.

In her early twenties Morgan was an English literature student but one who discerned no interesting career path ahead. She tried freelance journalism but found it poorly paid and just another nine-to-five. So, like many another twenty-somethings, she took a job in a bar-restaurant. The bar-restaurant happened to be the Shoreditch Electricity Showrooms in Hoxton Square, where Morgan found herself plunged willy-nilly into the roiling Brit art world.

She felt instantly at home though never saw herself as a functioning part of it. However, she was inspired to begin making things.

Happiness has been defined as the adult fulfilment of a childhood dream and many of Morgan's childhood energies had been centred on life and death in the animal kingdom. So, she would become a taxidermist.

She found a professional taxidermist in Edinburgh called George Jamieson, who offered to instruct her. She took a train, stayed with a friend, and was at Jamieson's early the following morning. "I spent a whole day with him, and did a pigeon; a very, very rushed job. He finished the thing off for me. I got a train back that night and was back at work the next day."

Then in February 2005 the Shoreditch Electricity Showrooms was sold. Morgan, who had a redundancy payment of £4,000, returned to Jamieson's Edinburgh workshop for a week.

It was around then that David Waddington, who had been one of her fellow bar managers, opened a new venture, Bistrotheque. He asked Morgan to make some work for four glass domes they had in the bar. The unveiling of the work turned into an event. "Loads of people turned up and lots of them were art world people I knew." She sold all four pieces to the bar for £350 each. "It was enough to keep me going for a little while. I was also working weekends at a café in Shoreditch where a friend was the chef, which paid about £150 and took care of my rent. Combined with my fortnightly dole cheque I could just about get by."

The next public appearance of Morgan's work was at the Zoo Art Fair in 2005, where *Rest a Little in the Lap of Life* was shown by the artist/dealer Wolfe Lenkiewicz. "Suddenly there was this big flurry of interest. Wolfe told me he could have sold that work twenty times over and that I should make more work. I didn't think I was good enough at taxidermy at that point but knew I had to seize the opportunity."

Among those who approached Morgan at Zoo was somebody who had talked with her about her work at Bistrotheque. She discovered this was Banksy. "He asked if I could make something for 'Santa's Ghetto', a yearly show he put on at Christmas time in a disused shop on or around Oxford Street. Then in 2006 Mollie Dent-Brocklehurst asked me to exhibit in her summer show at Sudeley Castle. I put three small works in what used to be the outdoor kennels. A year later, she organised 'The Exquisite Corpse', my first solo exhibition."

So Polly Morgan was well and truly launched. Her journey tells us something about the new art world, and its appetite for tapping into sources of visual energy way outside traditional parameters. This is an art world that can embrace the aerialist Philippe Petit, the photographer David LaChapelle and such makers of theatrical movies as David Lynch, Agnès Varda and Wim Wenders. It was more than ready for Polly Morgan.

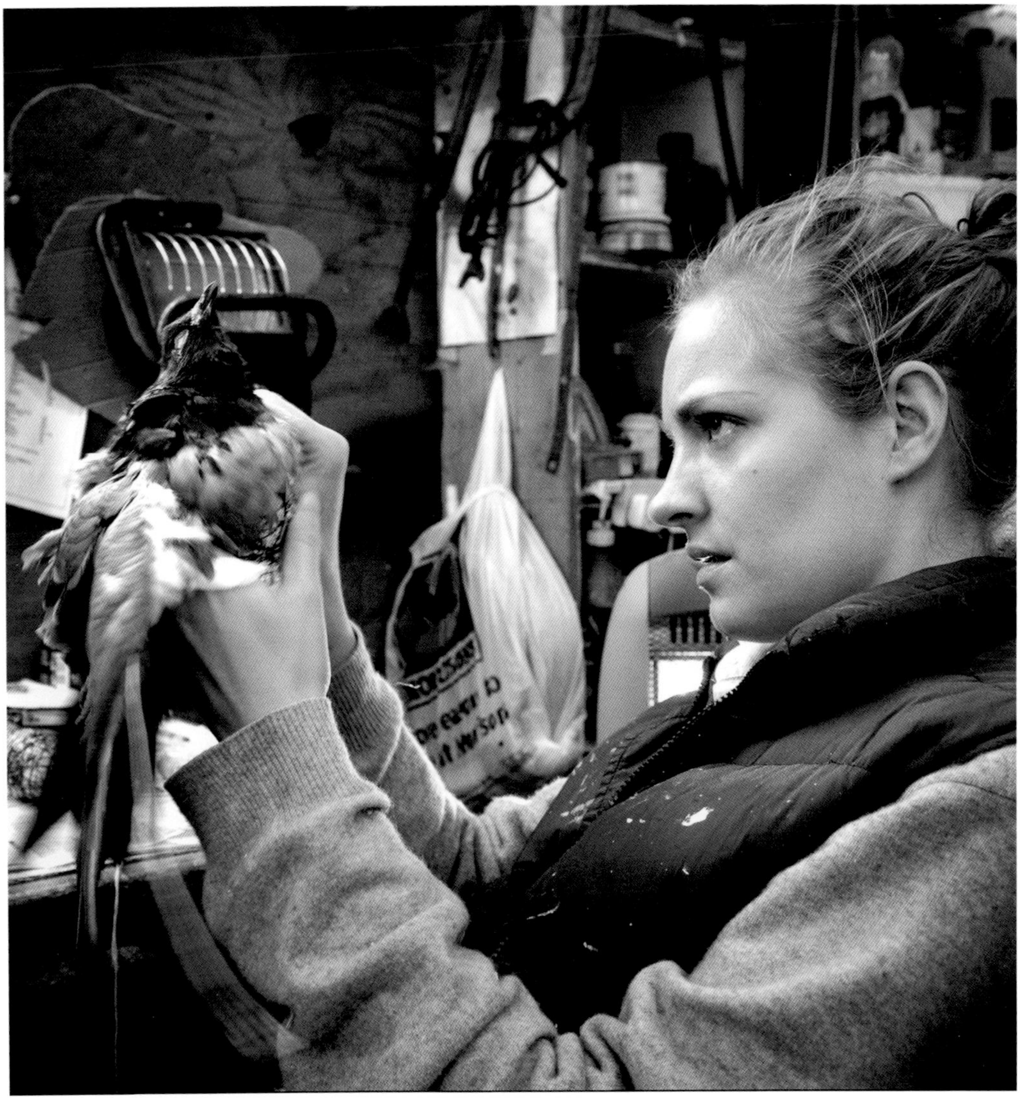

Back to Go.

"I started off learning taxidermy with no particular intention to become an artist but neither did I want to be a traditional taxidermist, mounting pets or hunting trophies", nor part of the 'creative' tradition which was a sort of nineteenth century proto-surrealism of which the most famous practitioner was Walter Potter.

"His taxidermy wasn't technically skilled," Morgan says. "But he became quite legendary because he made these kitsch, outlandish displays; a kitten's wedding where they are all dressed in finery and a tea party where they are all huddled around a table, drinking out of little china cups. He would also taxidermy freak animals, such as chickens with four legs or lambs with two heads." An outcrop of this were the so-called 'Rogue Taxidermists' who would use animal parts to make beasts-that-never-were, like dragons or mermaids.

"When I first started learning, almost everything I found was in a naturalistic setting, with the animal looking as it had in life. I wanted to keep the birds or mammals looking dead, as they looked when they came to me. I can understand why people want to resurrect them, but I thought there was something quite beautiful about their death pose that I thought worth preserving."

The first works that Morgan made after Zoo, the works that were shown in 'The Exquisite Corpse', for instance, were relatively uncomplicated. They derive their energy from the sheer joy that Morgan found in the process, so for a couple of years, in her phrase, she "buffered along". But one of the salient characteristics of the art world is that it is an information system and those who work within it learn to play by its rules. Healthy organisms need to reach, to grasp, to grow. Two recent pieces tell the story: *Departures* and *Carrion Call.*

"I wanted my work to become less ornamental and more monumental. I started to be more thoughtful about what I undertook and not to rush into anything," she says. "Holding onto the idea for a while and letting it develop in my mind before getting started."

Departures is based on an illustration Morgan found in a book about Victorian inventions. It is a steel engraving which illustrates a letter to The Scientific American proposing a flying machine to be powered by birds.

"I wanted to build it as an homage to the inventor, who clearly never realised his plan. So I redesigned it and made the carriage like a giant birdcage. I like the idea that it's not quite clear whether the passenger or the birds are free. The passenger has the freedom to travel while the birds are tethered, yet they have the real ability to fly while he is caged. I like that dichotomy between the human and the animal. We harness them in all kinds of ways but they are the ones with ultimate power."

In the steel engraving the birds are eagles while Morgan used three vultures and a bouquet of other coloured birds.

She worked on it on and off for about a year and showed it in October 2009 at a show called 'The Age of the Marvellous'.

Carrion Call is a large wooden coffin made from old distressed wood and stained to look as if it was rotten. Out of holes appear clusters of tiny quail chicks. As with the birds who are the slaves/masters of the flying machine there is a dichotomy here. We are conditioned to see young birds, indeed the young of most animals, as loveable, fluffy little packages of life-force. But little birds are fearsome too; tiny engines powered by hunger and need, and the chicks in *Carrion Call* explode from the innards of the coffin like maggots from the intestinal tract of an exhumed corpse. It's a vibrant piece, but cold-eyed and scary too.

"I had been working with quail chicks on and off for years. I've done a few versions of *Carrion Call.* In one, *Pyric Victors*, the chicks were dyed black to look like ash and the coffin had been set alight and burned, like the phoenix rising from the flames. With the work involving chicks, I can see very clearly the way it has developed. The first was tiny, just one chick curled up inside a tiny matchbox." Like one of the burials she had made as a child.

- - - -

Just when this text was almost done Polly Morgan told me about a curious dream she had just had. It featured animals, mostly birds, was sequential and somehow related to the months of the year. It was a bit blurry and she didn't understand it at all. But perhaps one day this too will emerge in a flurry of wings and open mouths.

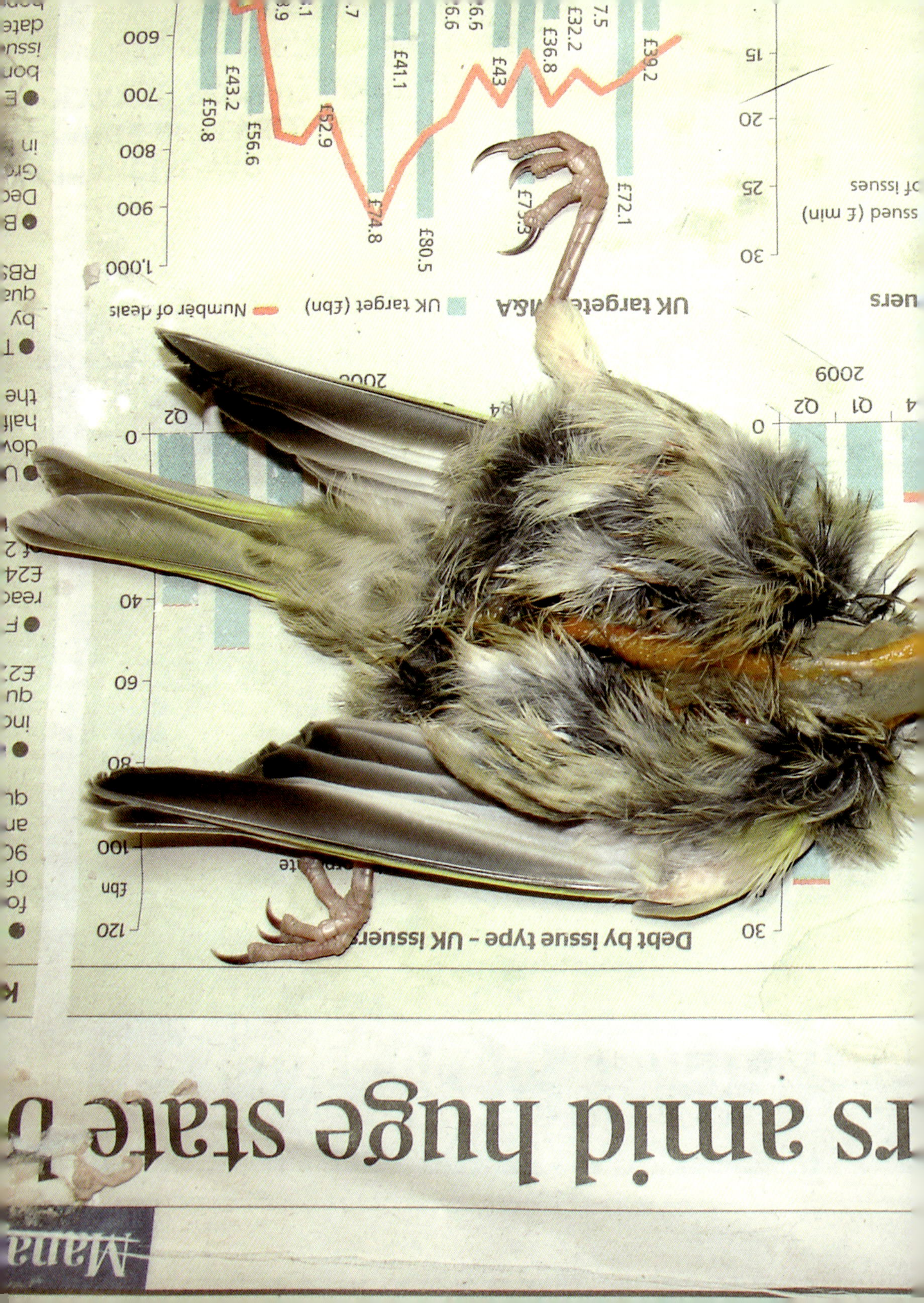
rs amid huge state
Debt by issue type – UK issuers
UK targeted M&A
UK target (£bn)
Number of deals

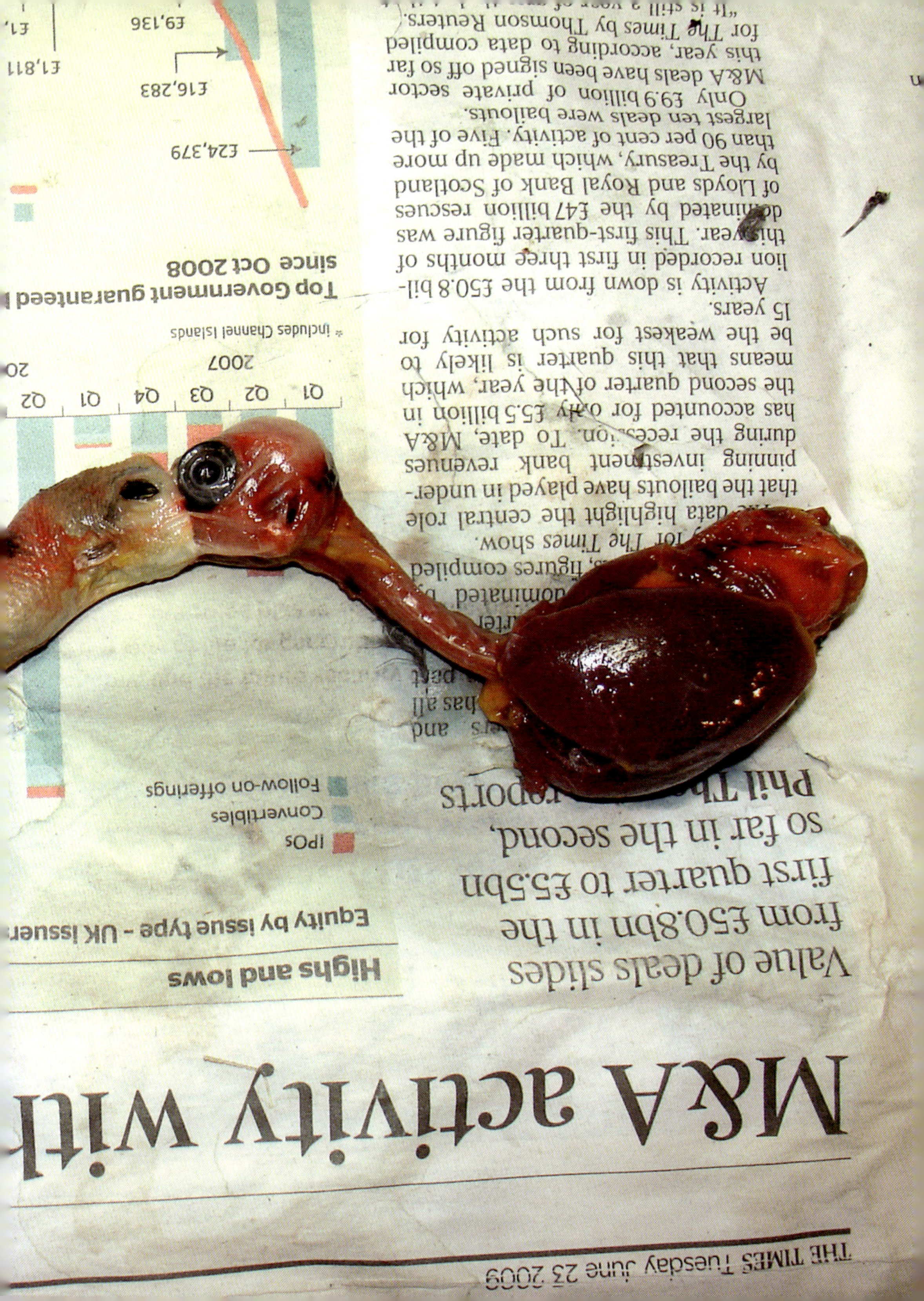
THE TIMES Tuesday June 23 2009
M&A activity wit
Value of deals slides from £50.8bn in the first quarter to £5.5bn so far in the second,
Phil Th reports
ers and
has all
part
arter
dominated by
, figures compiled
for The Times show.
data highlight the central role that the bailouts have played in underpinning investment bank revenues during the recession. To date, M&A has accounted for only £5.5 billion in the second quarter of the year, which means that this quarter is likely to be the weakest for such activity for 15 years.
Activity is down from the £50.8 billion recorded in first three months of this year. This first-quarter figure was dominated by the £47 billion rescues of Lloyds and Royal Bank of Scotland by the Treasury, which made up more than 90 per cent of activity. Five of the largest ten deals were bailouts.
Only £9.9 billion of private sector M&A deals have been signed off so far this year, according to data compiled for The Times by Thomson Reuters.
Highs and lows
Equity by issue type - UK issuer
IPOs
Convertibles
Follow-on offerings
Q1
Q2
Q3
Q4
Q1
Q2
2007
* includes Channel Islands
Top Government guaranteed
since Oct 2008
£24,379
£16,283
£9,136
£1,811

Polly Morgan

= =

Selected Works

- -

2005 - - 2010

Black Fever

2010

Taxidermy crow wings
Wood
Wire

116 × 83 × 83 cm

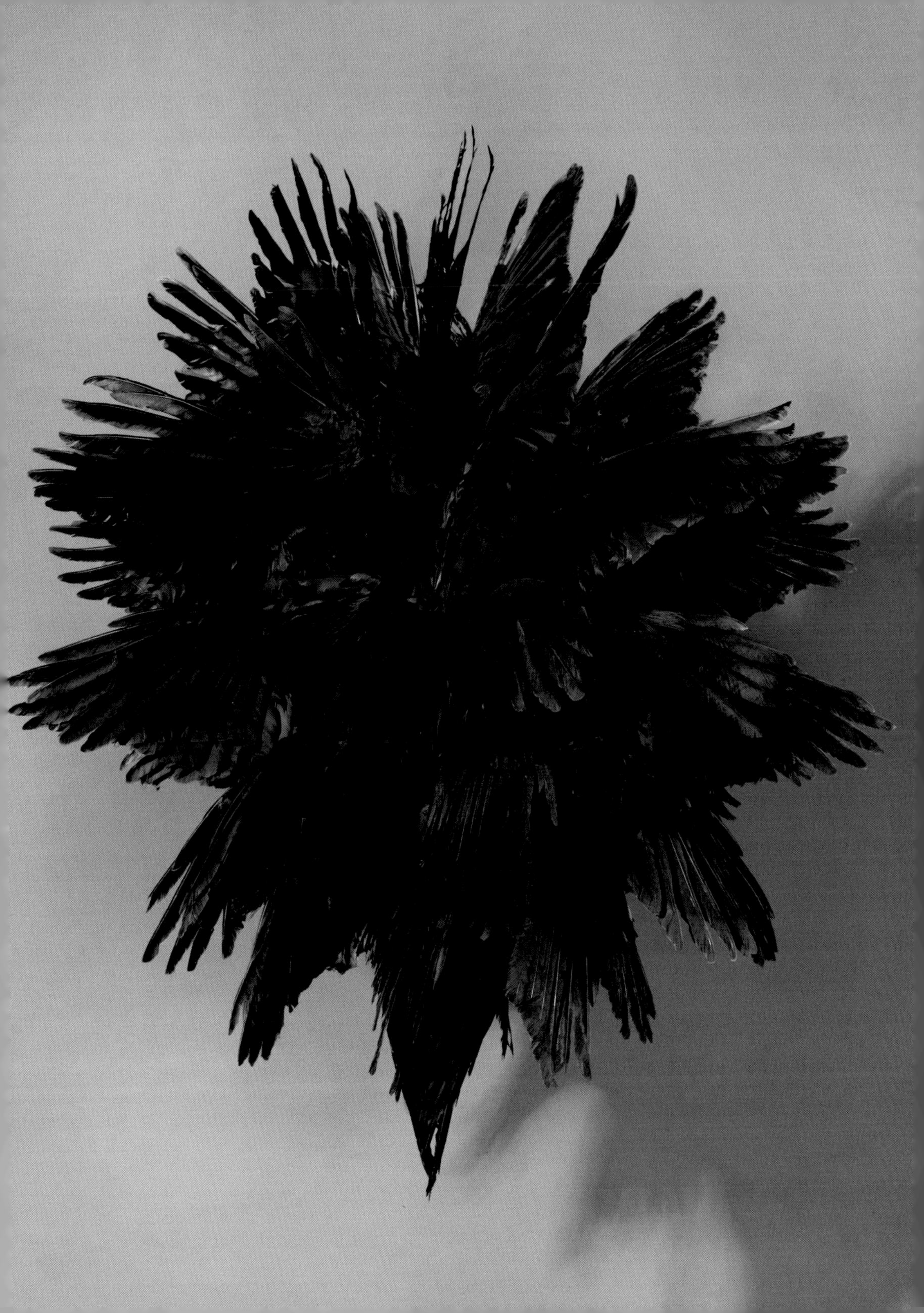

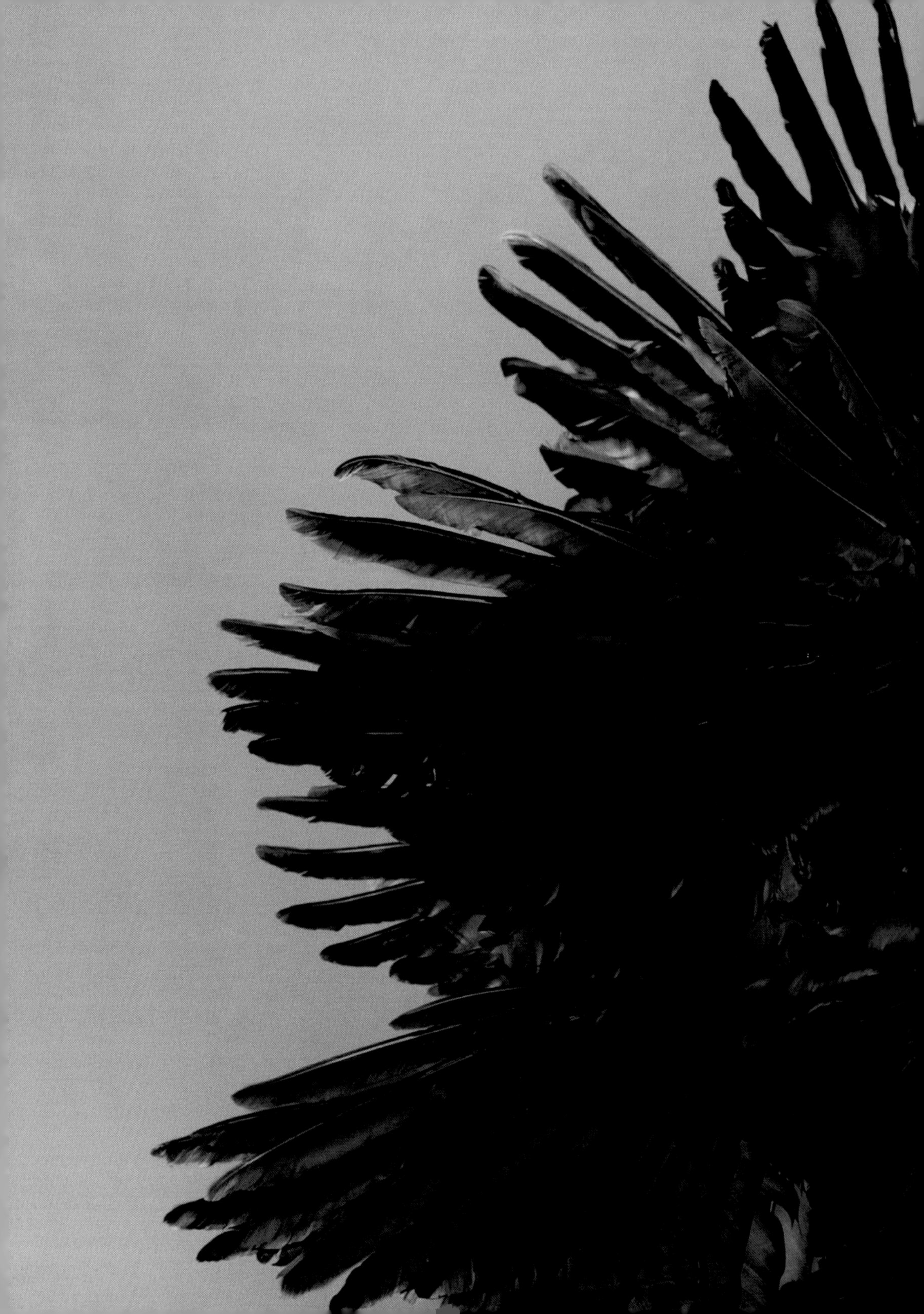

Blue Fever

2010

Taxidermy pigeon wings
Wood
Wire

120 × 80 × 80 cm

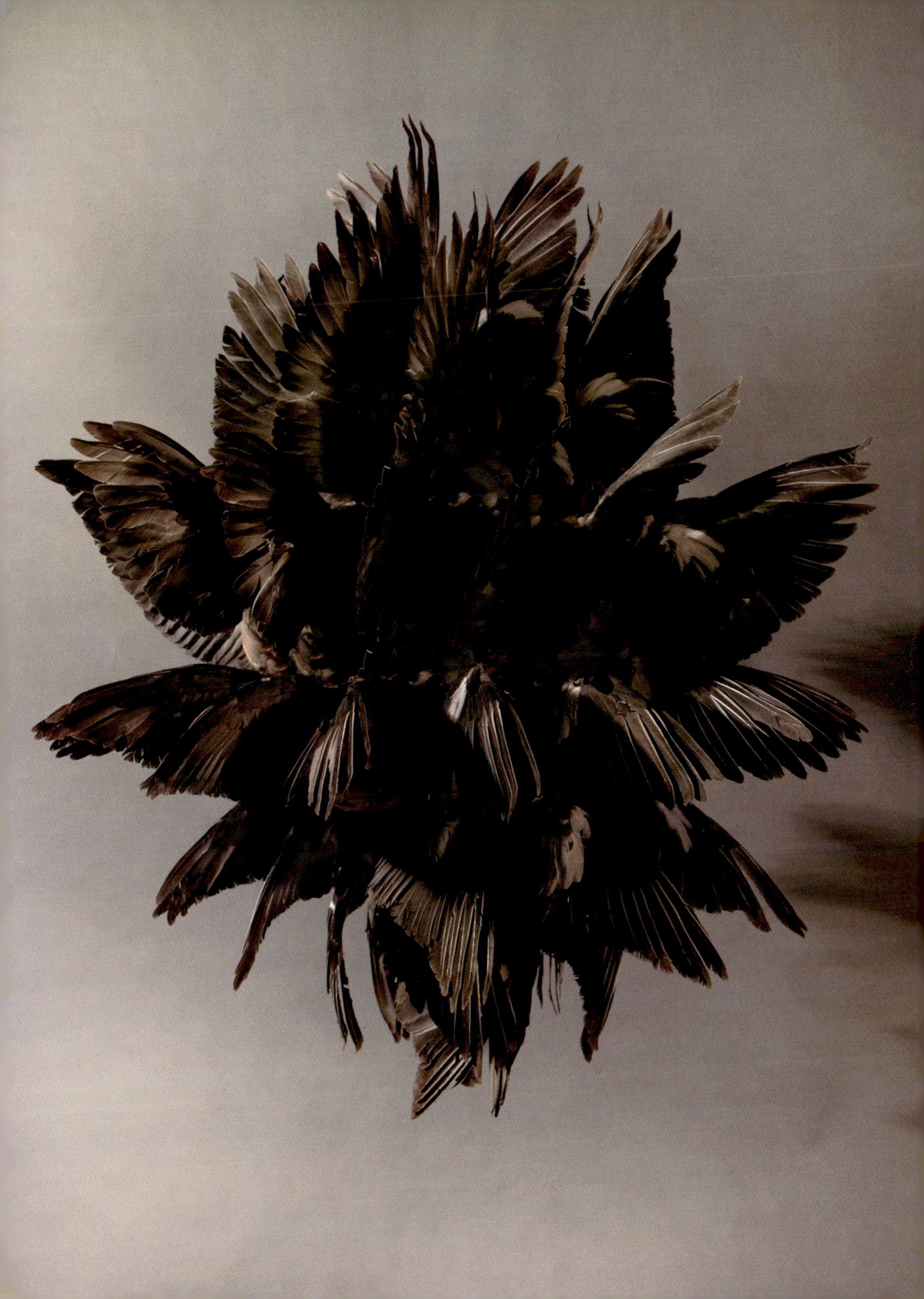

Systemic Inflammation

2010

Taxidermy finches
Taxidermy canaries
Steel
Leather

130 × 113 cm diameter

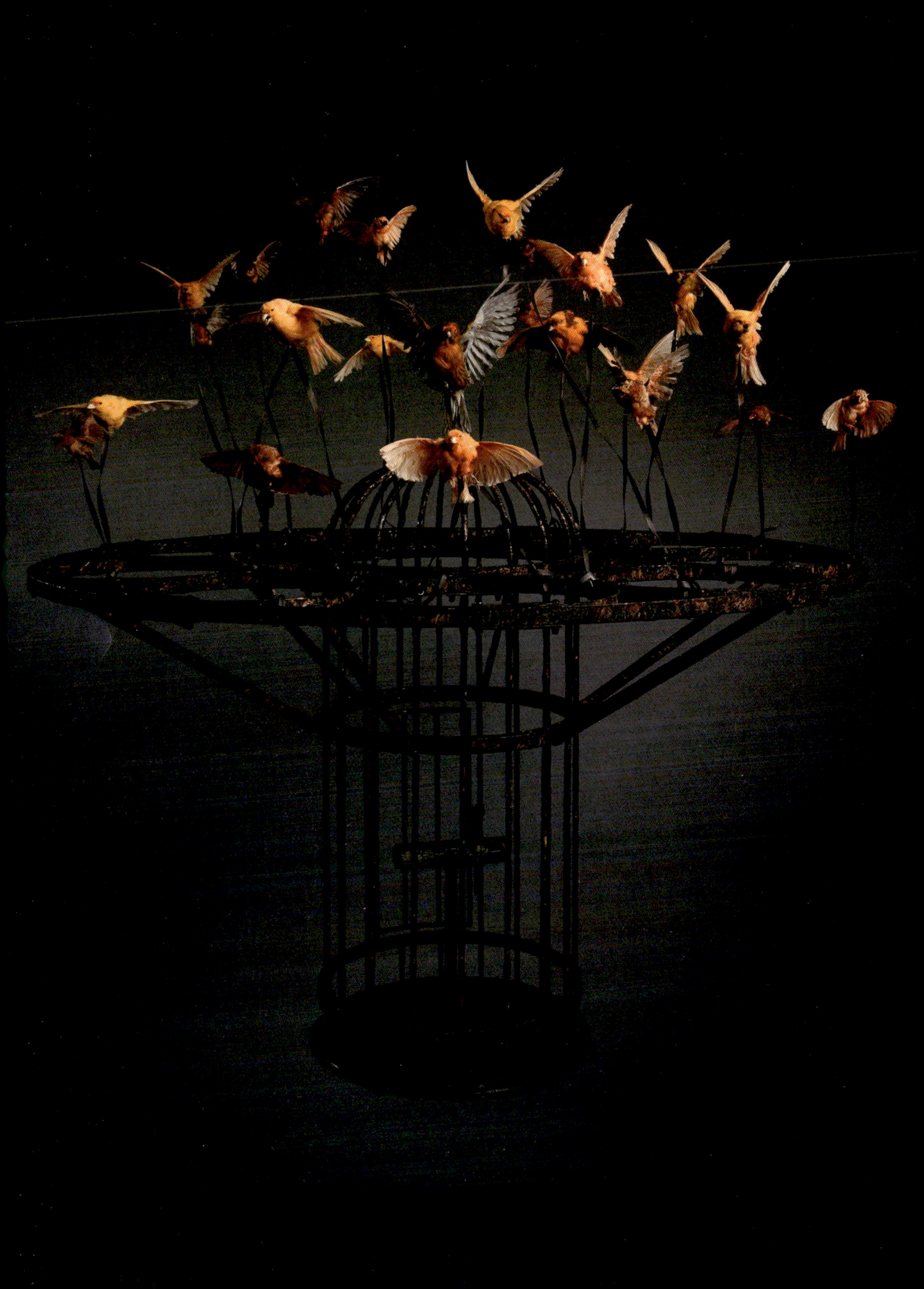

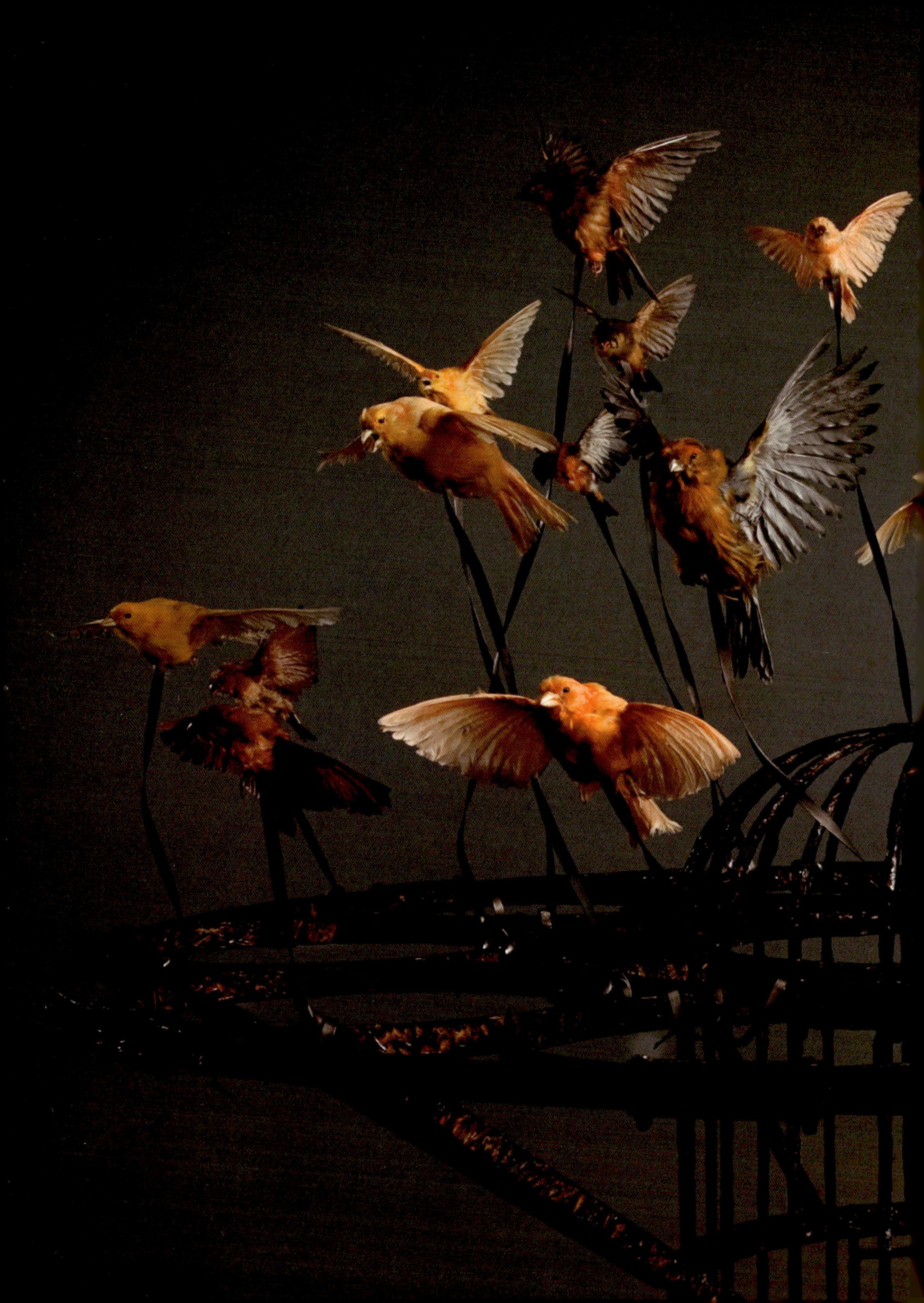

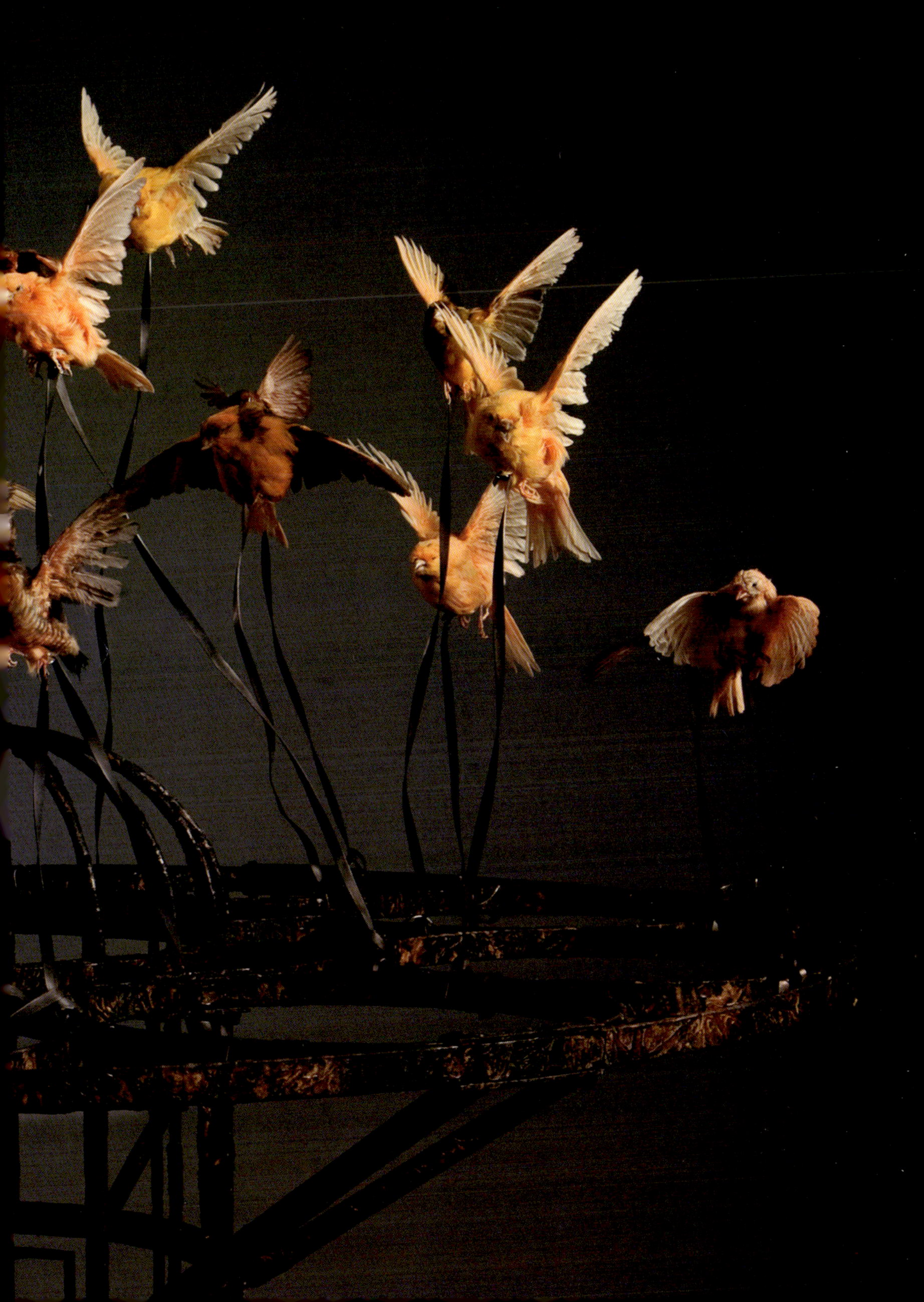

Atrial Flutter

= =

2010

- -

Taxidermy cardinal
Resin
Plastic
Wire

- -

85 × 27 × 22 cm

- -

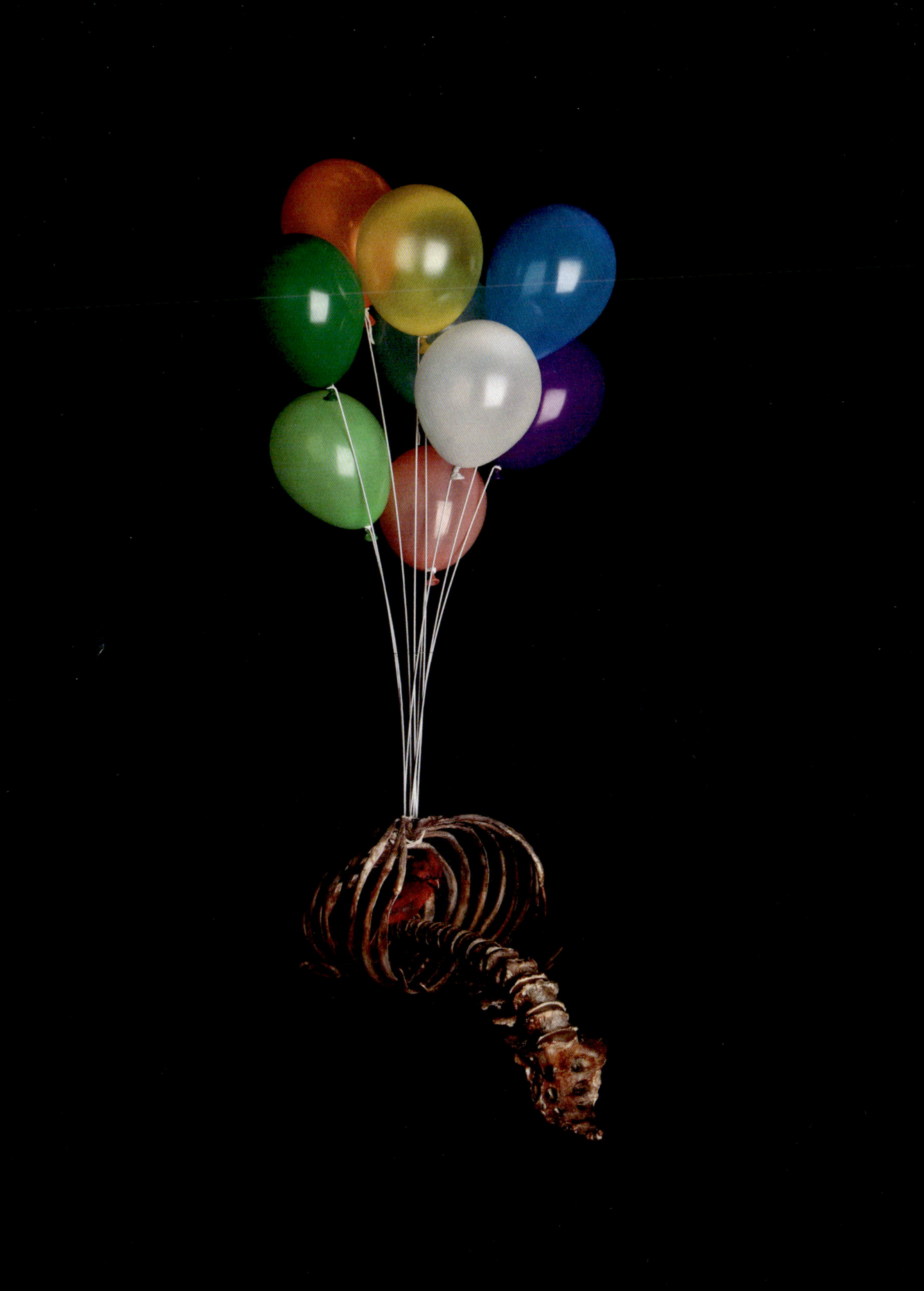

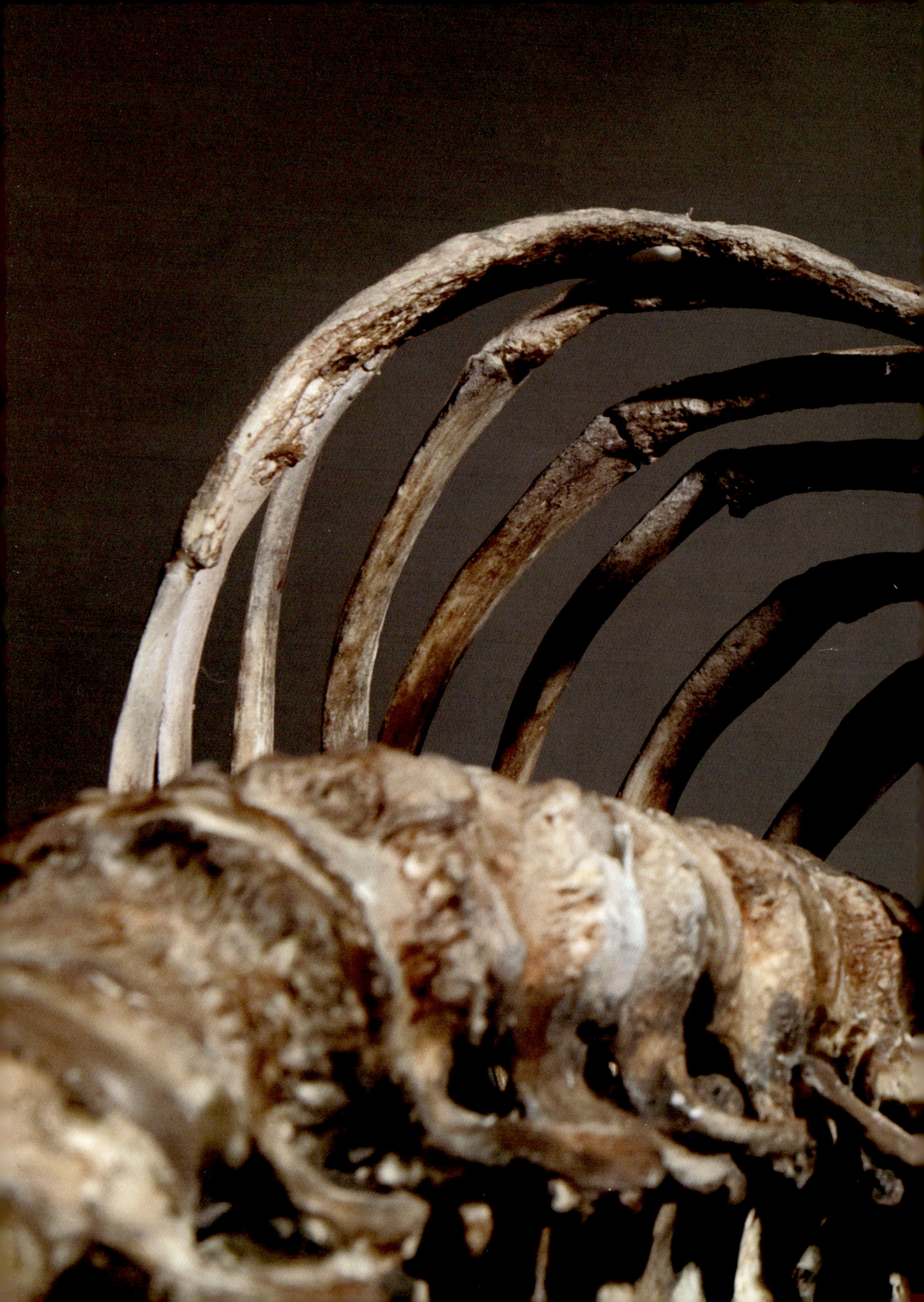

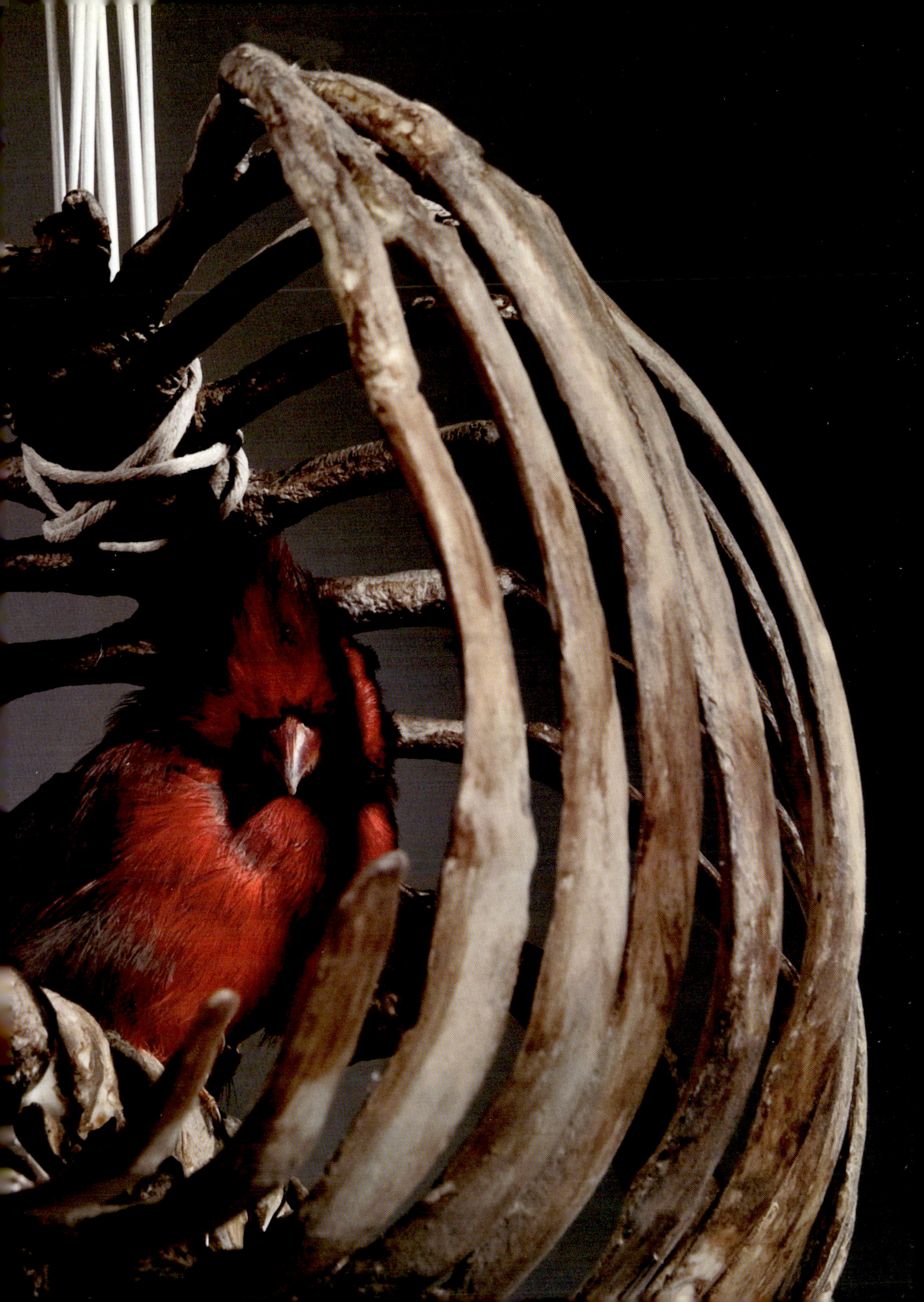

Still Birth

= =

2010

- -

Taxidermy pheasant chick
Wire
Resin
Glass
MDF

- -

52 × 26 cm diameter

- -

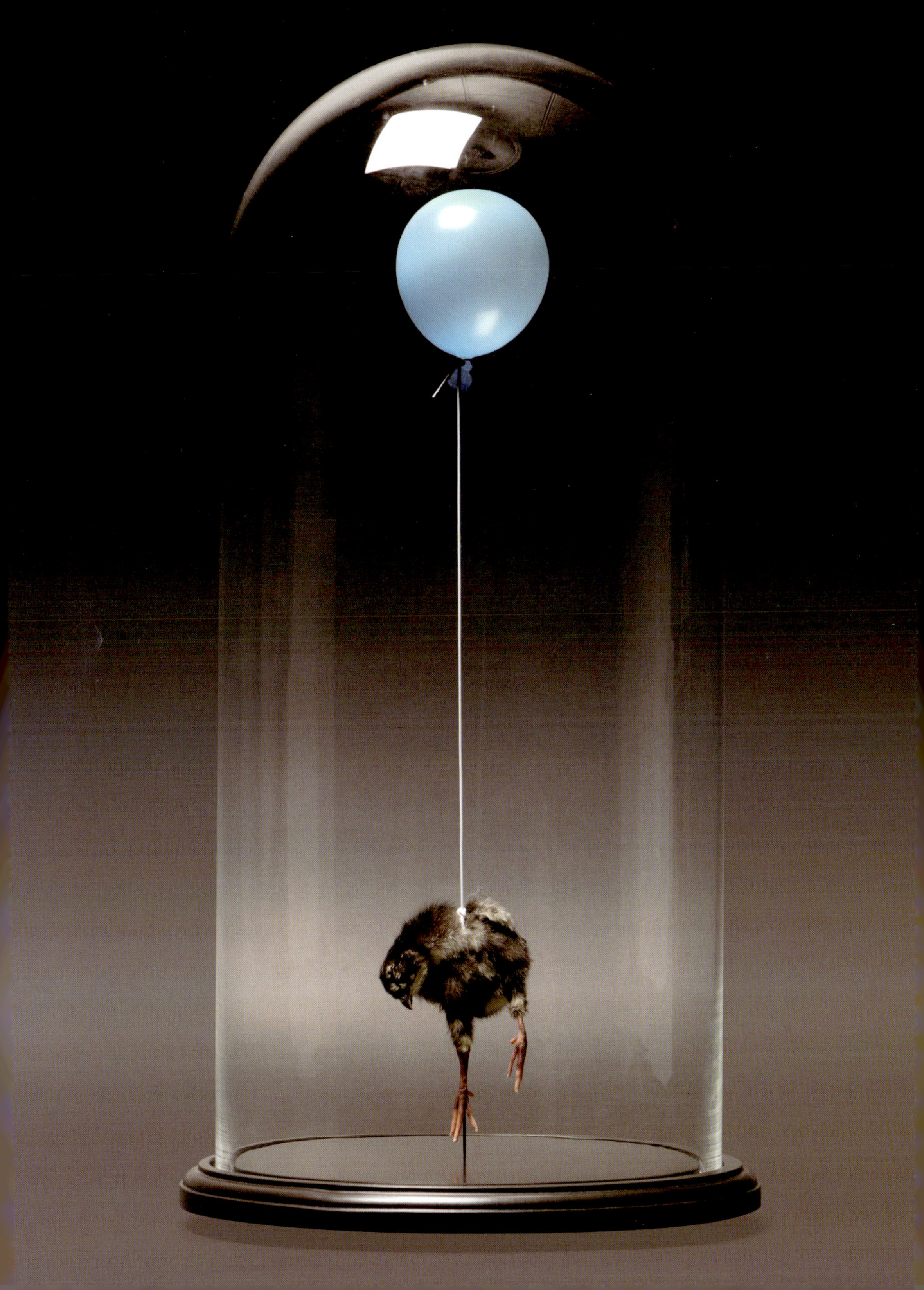

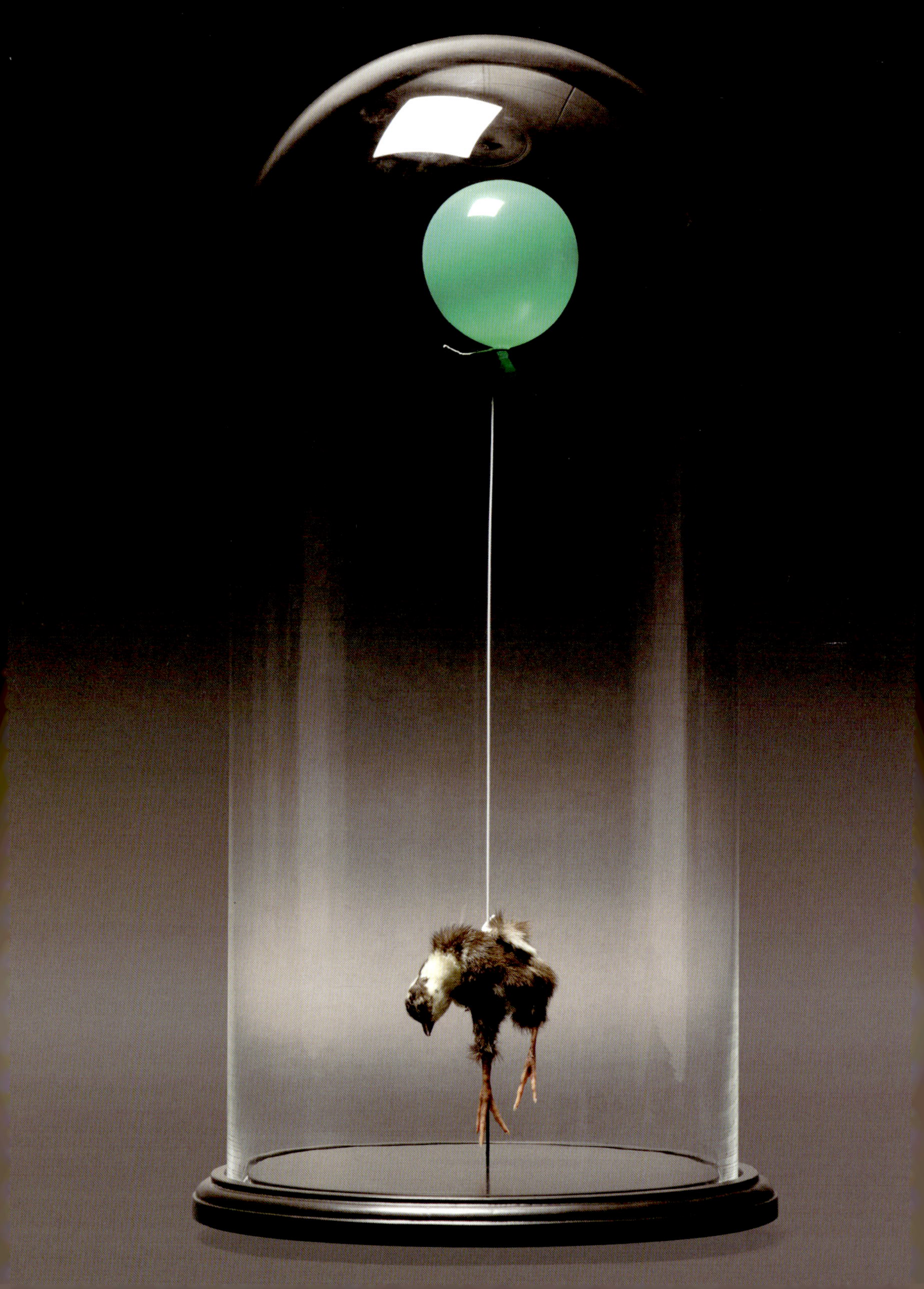

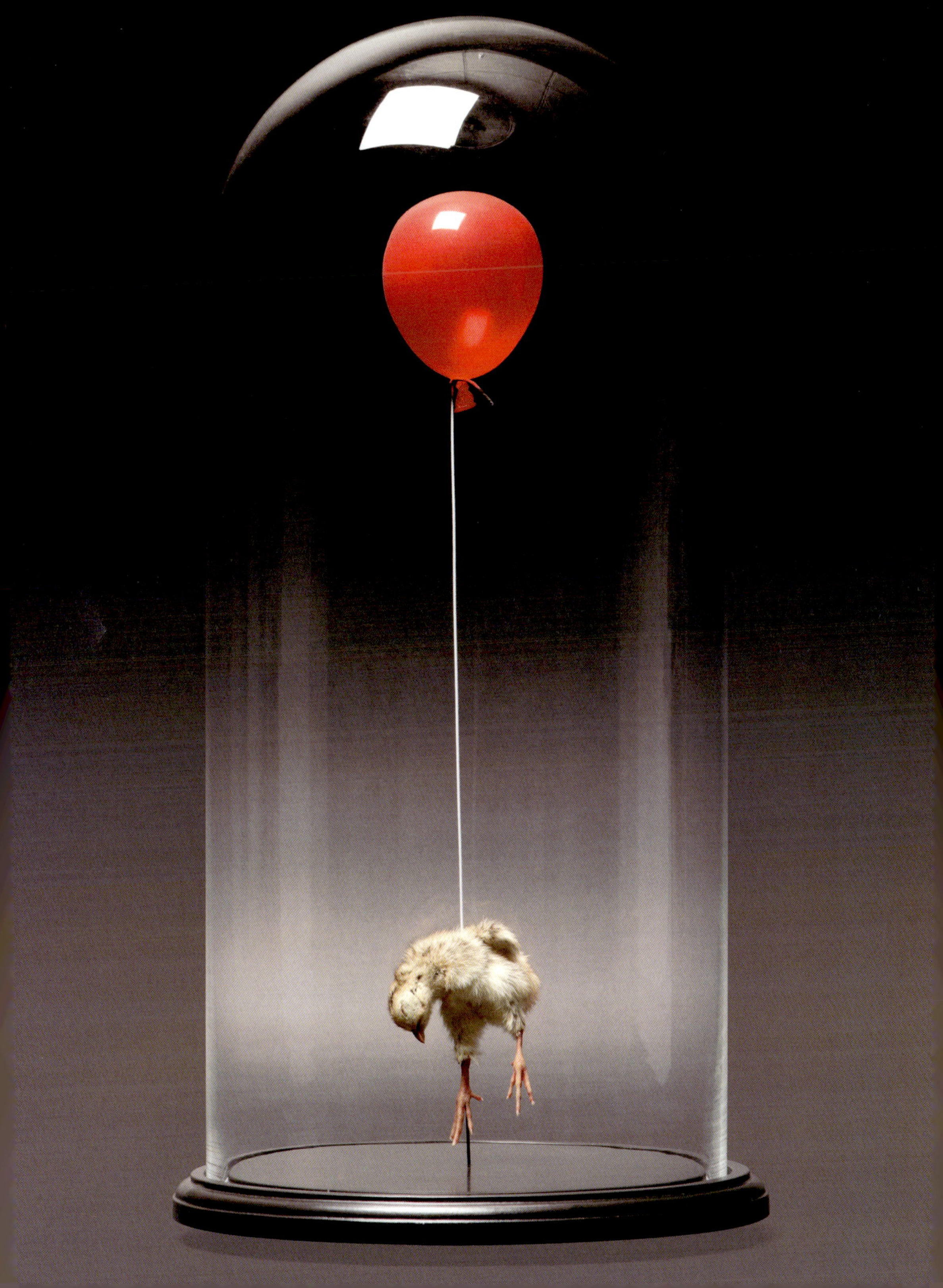

Feed Me

2010

Taxidermy bird heads
Wire
Metal
Resin
Lacquered bronze

41 × 17 cm diameter

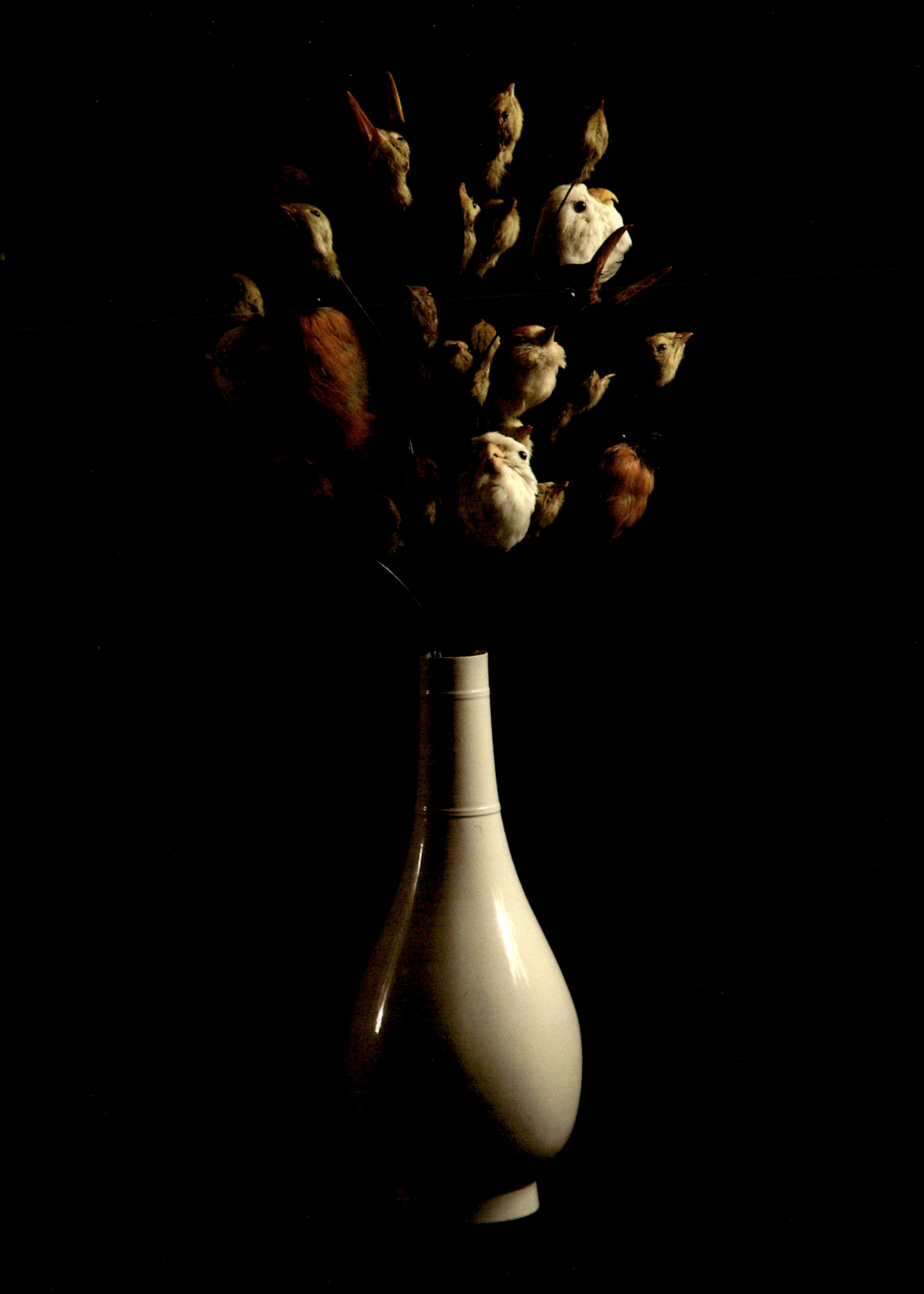

Departures

= =

2009

- -

Taxidermy birds
Leather
Brass
Wood

- -

400 × 250 cm diameter

- -

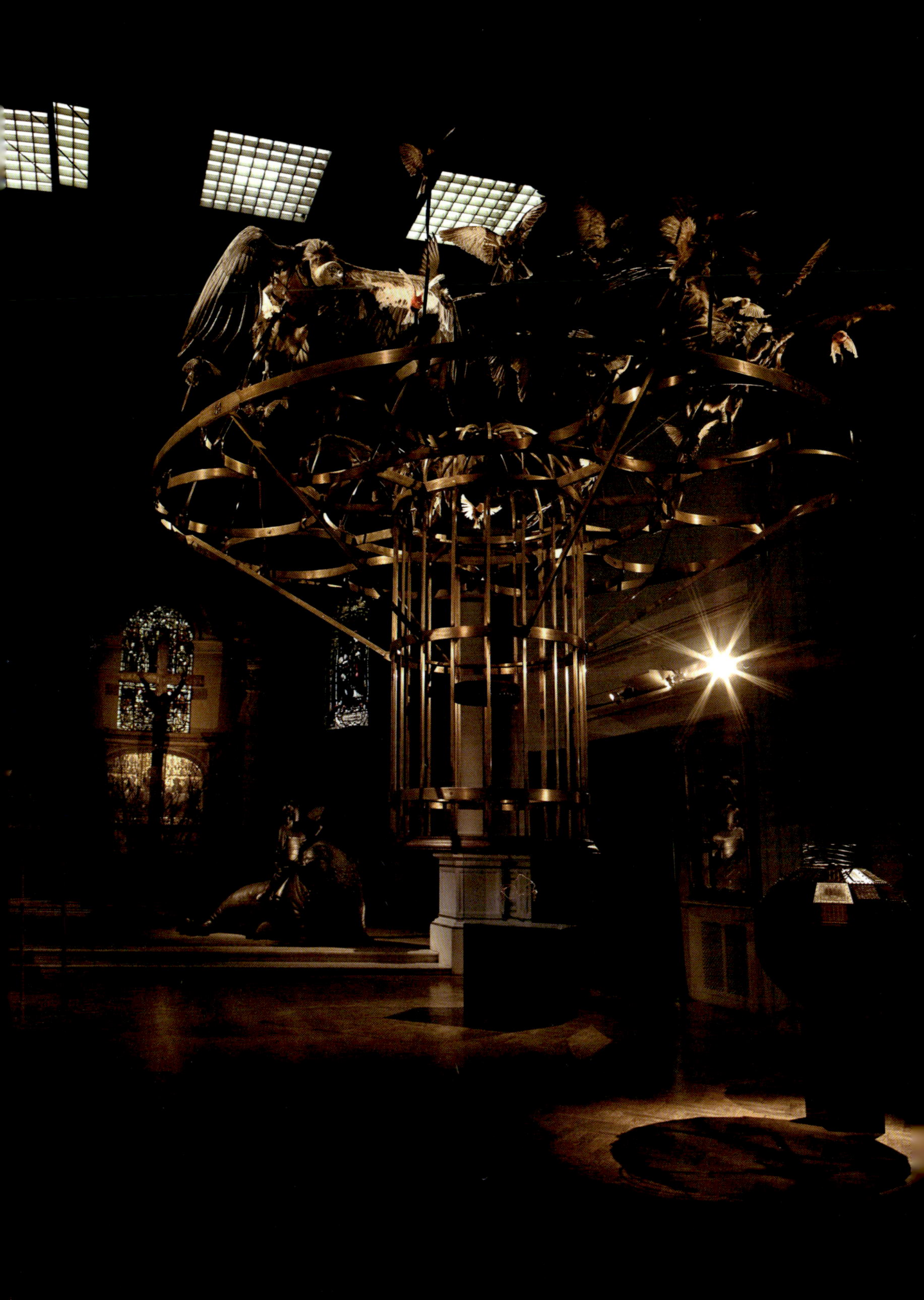

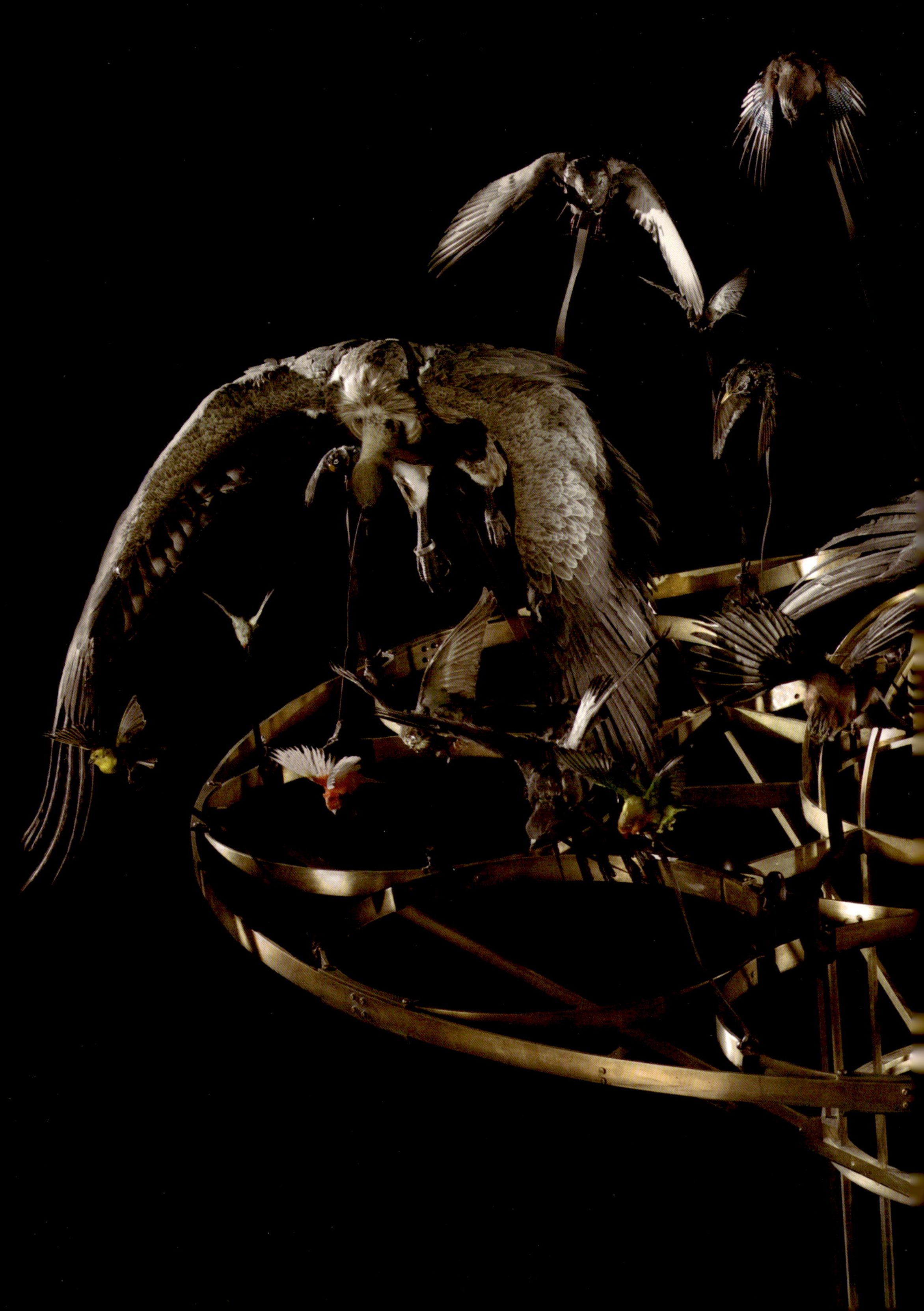

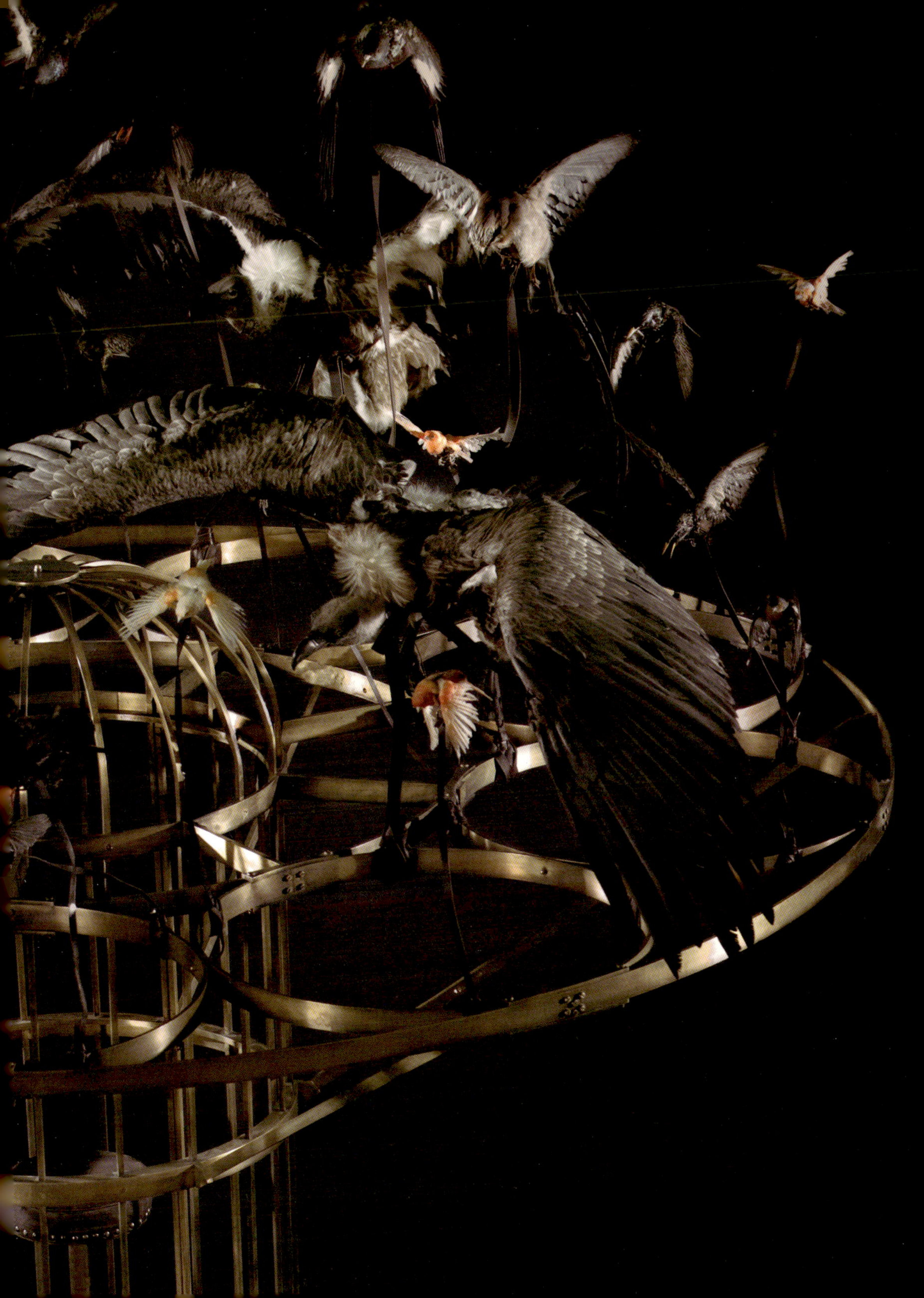

Carrion Call

2009

Taxidermy quail chicks
Wood

37 × 61 × 186 cm

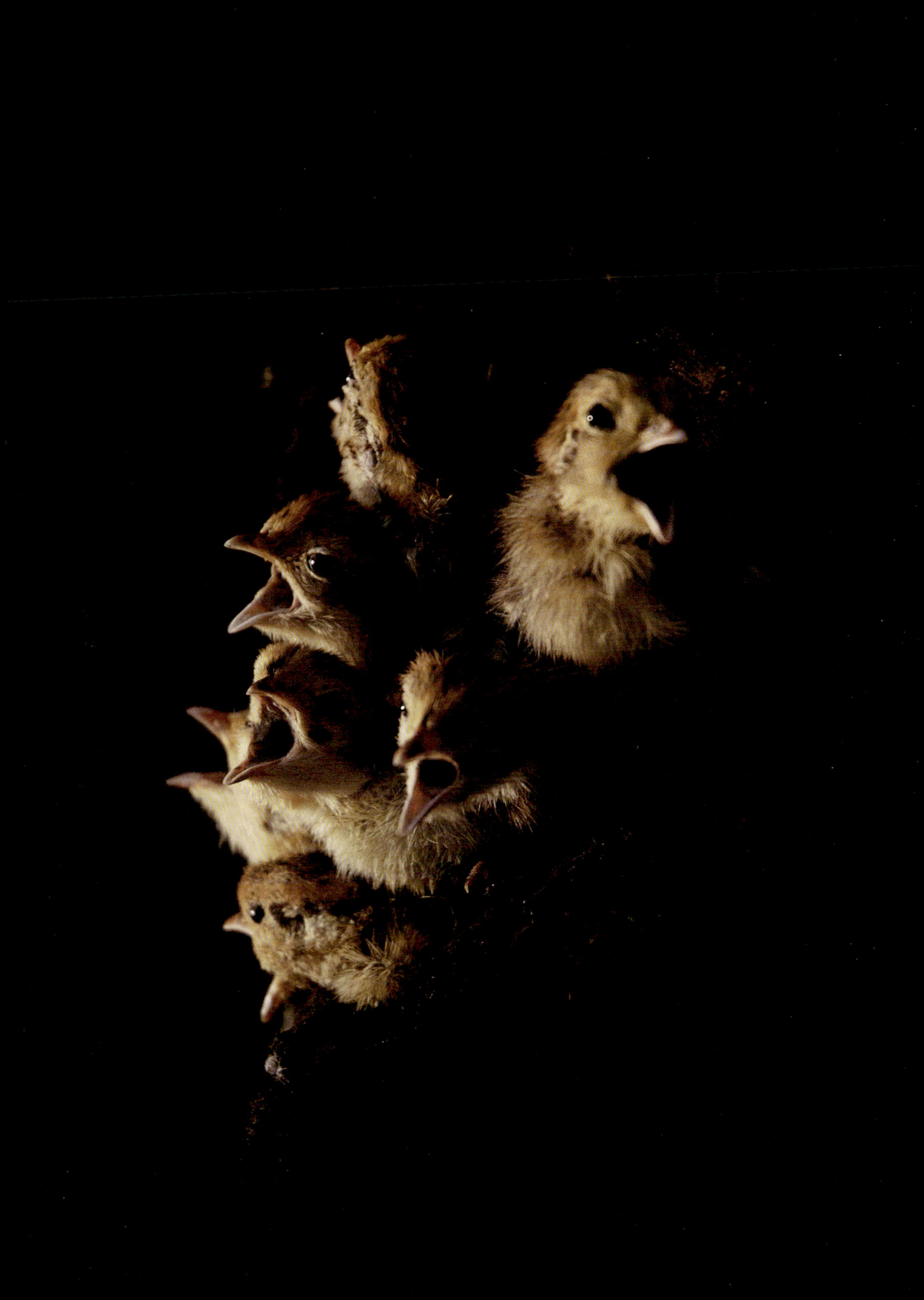

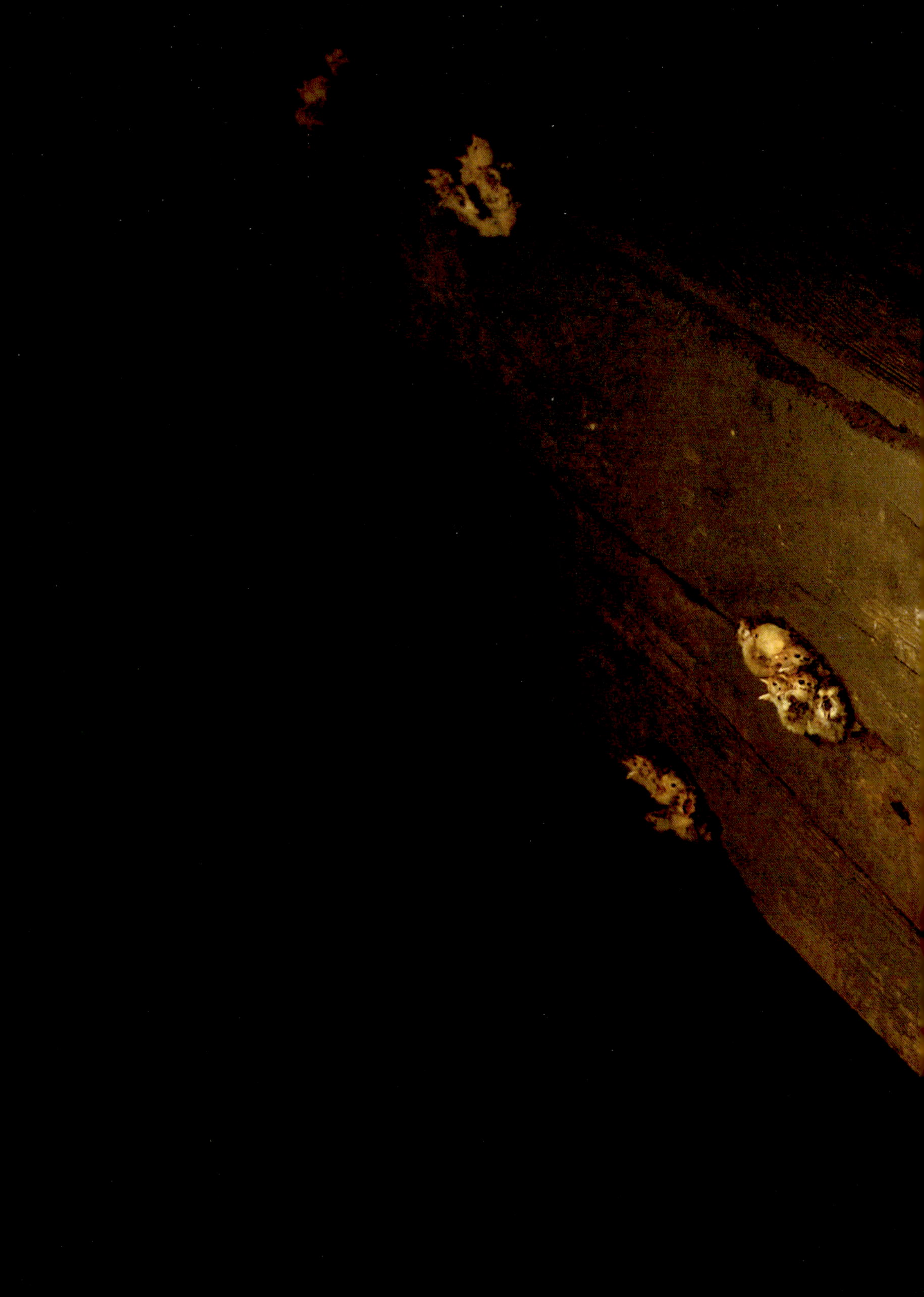

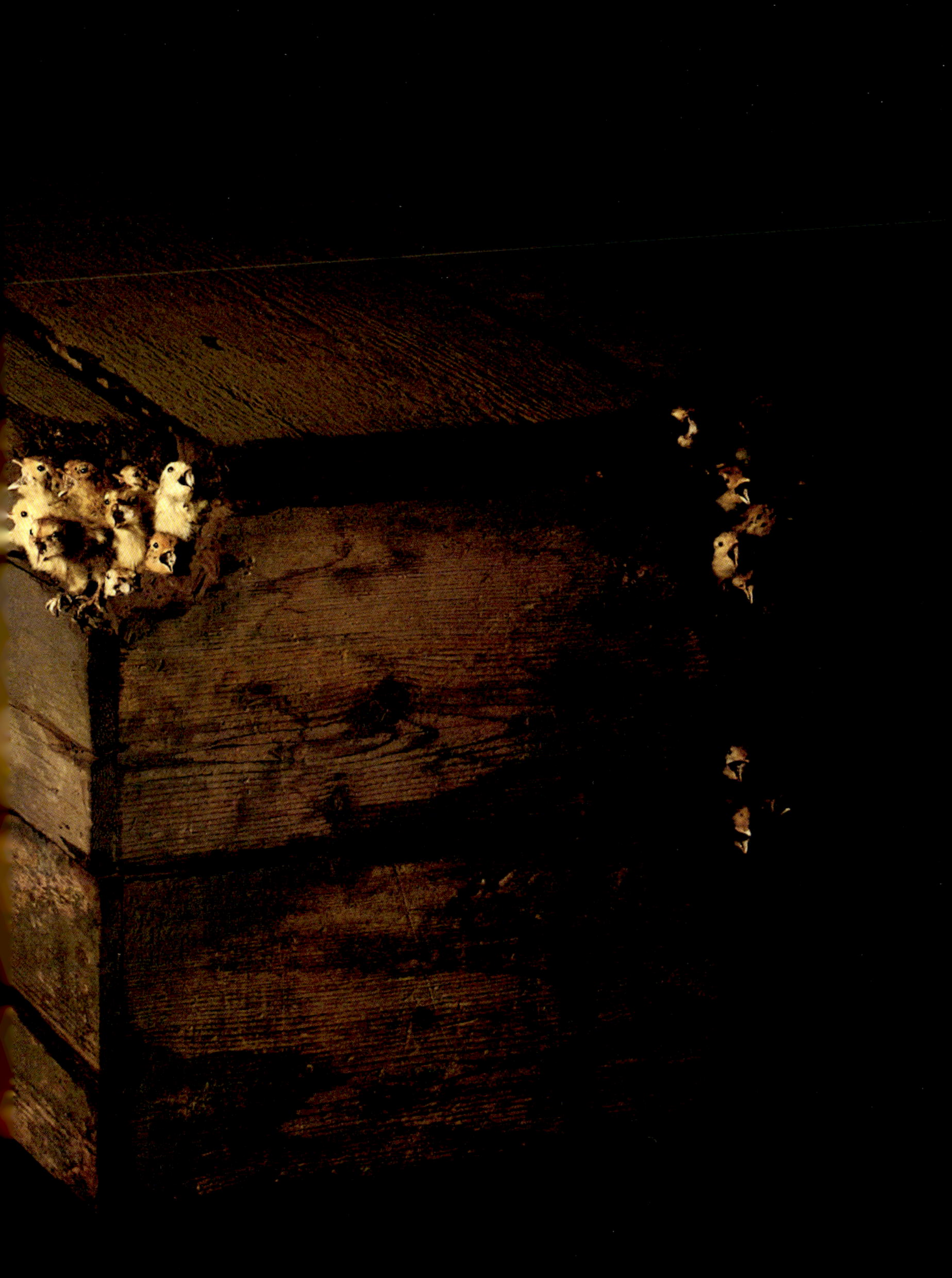

Vestige

2009

Taxidermy kingfisher
Leather-bound prayer book
Crystal chandelier
Glass
Wood

37 × 58 × 19 cm

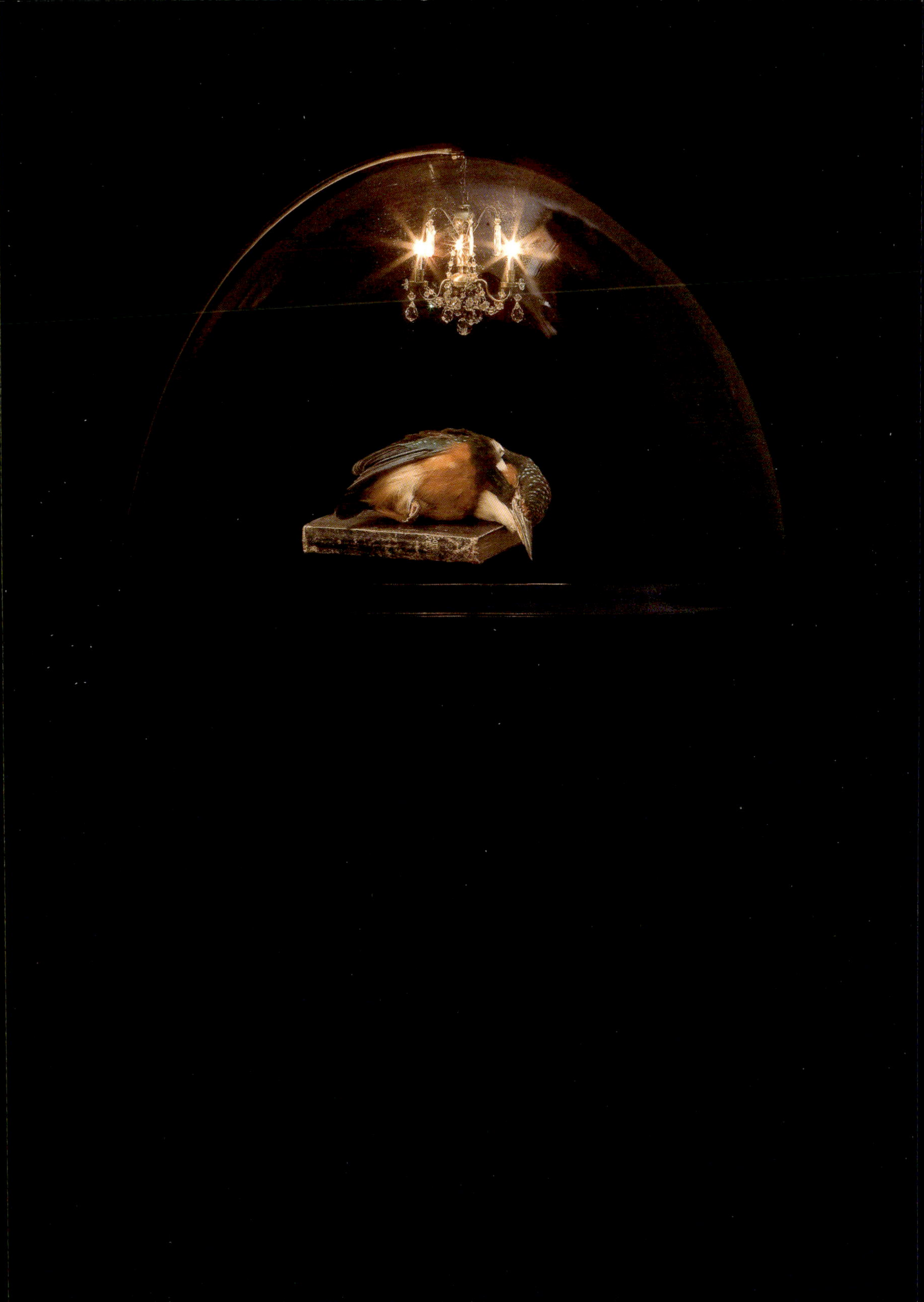

Flight of Fancy

= =

2009

- -

Taxidermy nuthatch
Crystal jewellery box

- -

9.6 × 14 × 9 cm

- -

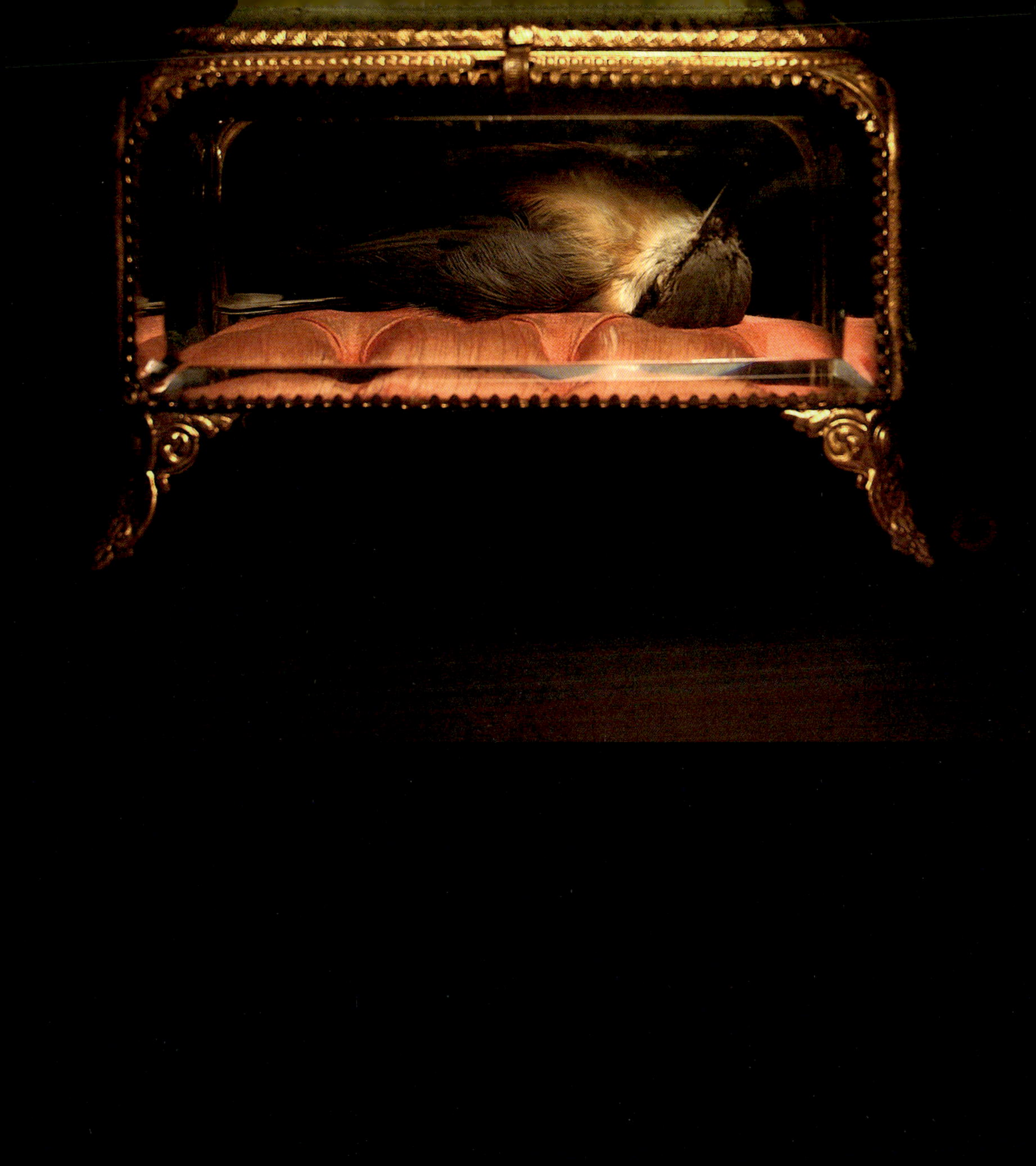

Dead Ringer

= =

2009

- -

Taxidermy magpie
Bakelite phone

- -

25 × 36 × 17 cm

- -

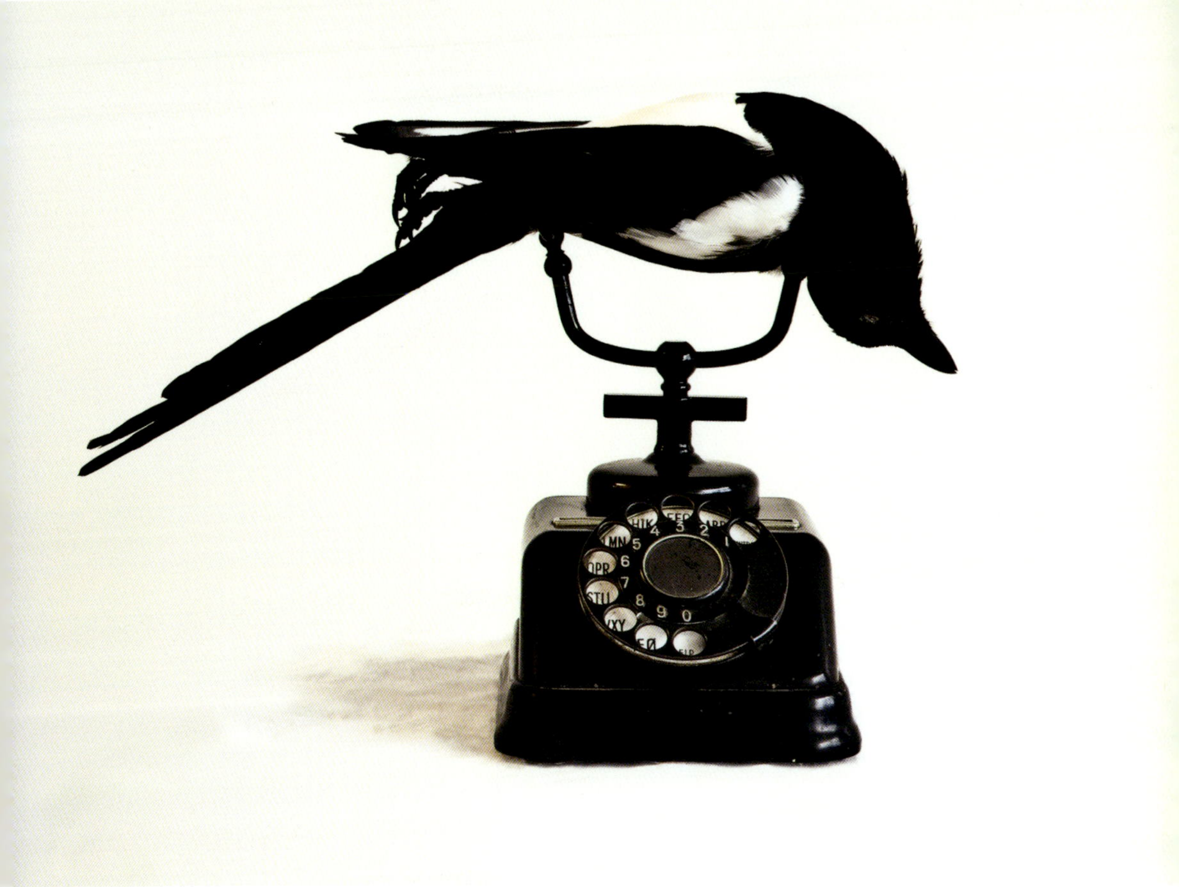

Receiver

= =

2009

- -

Taxidermy quail chicks
Bakelite handset

- -

23 × 9 × 6 cm

- -

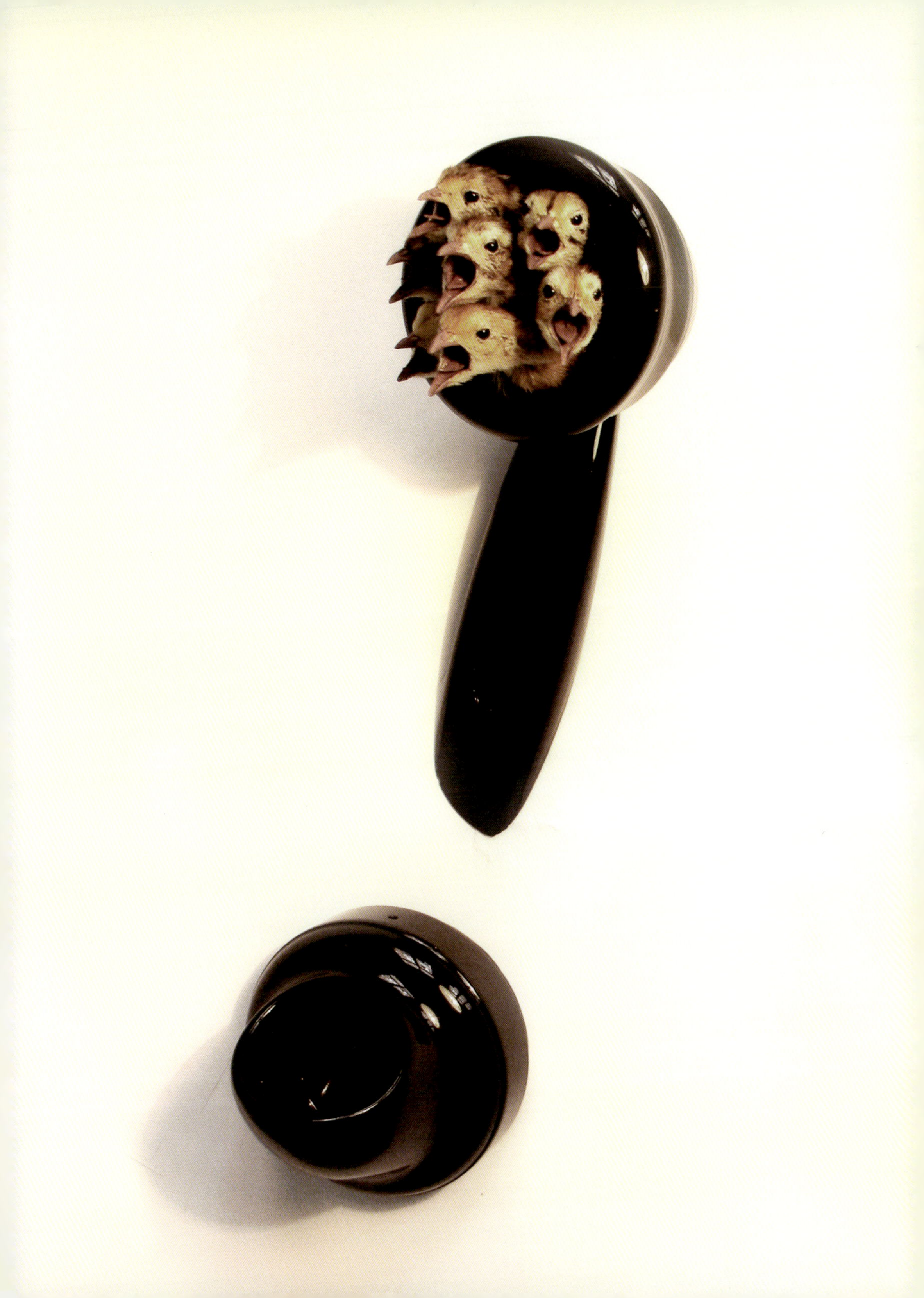

Pyric Victors

2009

Taxidermy chicks
Wood
Silicone
Resin

17.8 × 32.6 × 90 cm

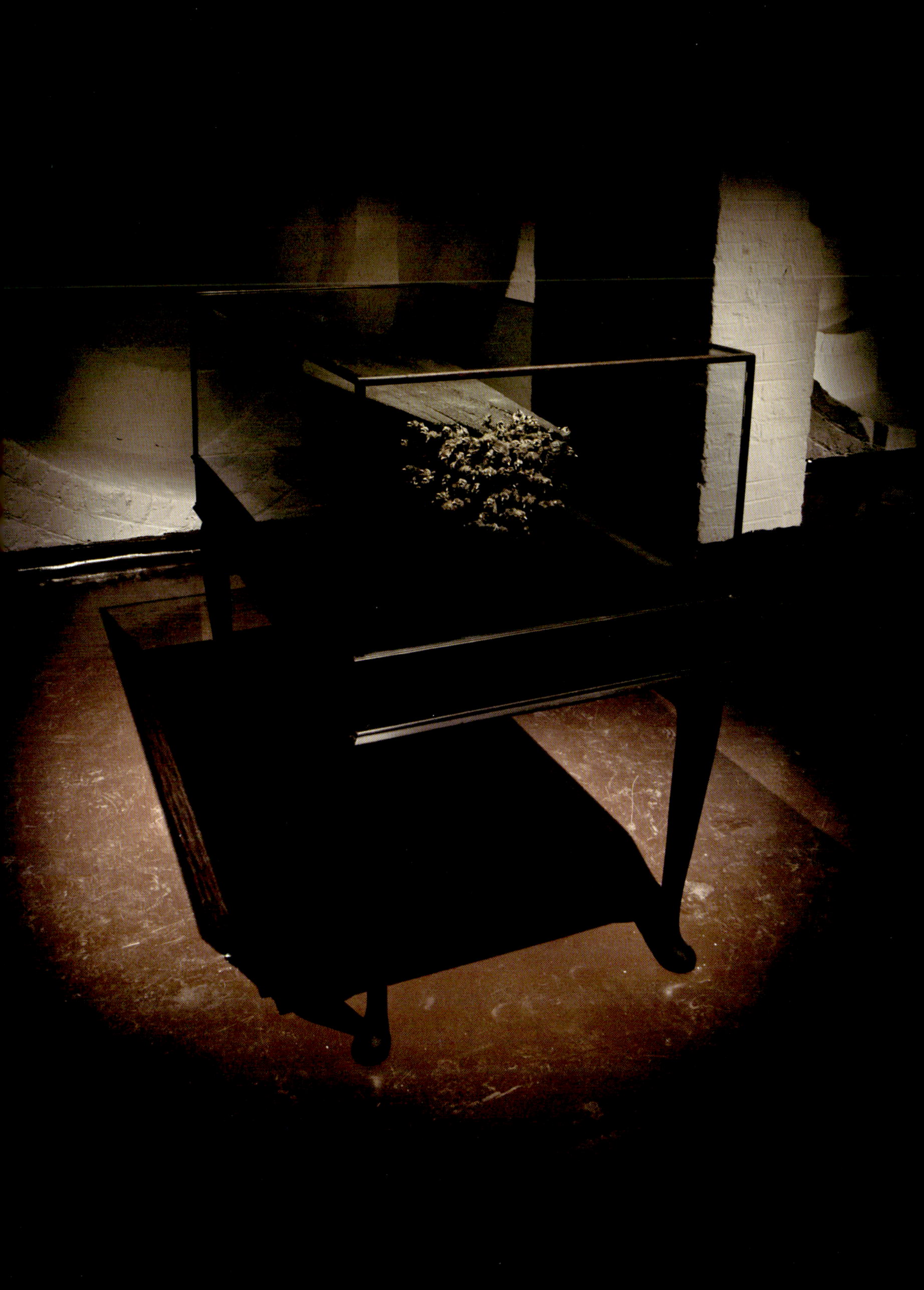

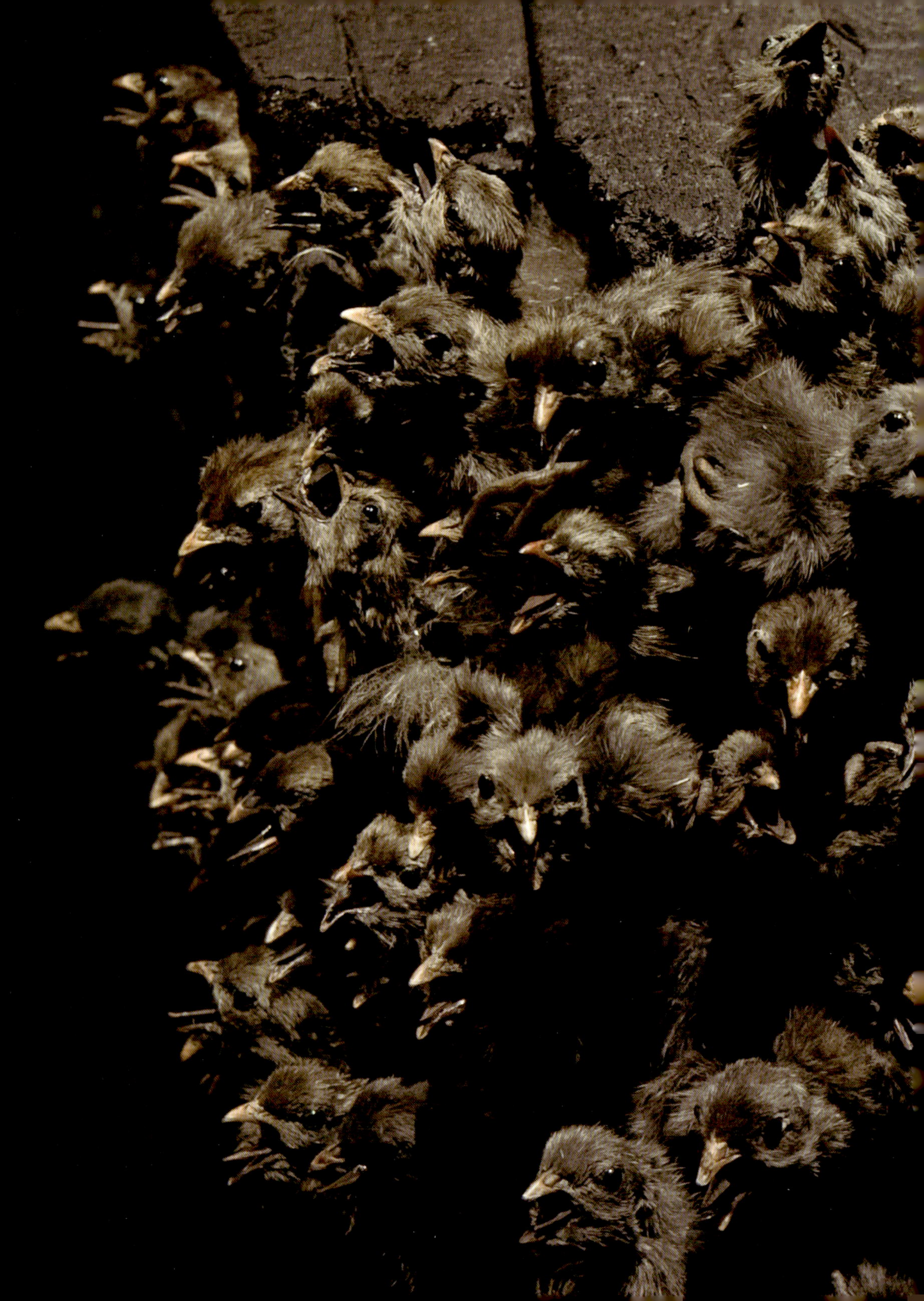

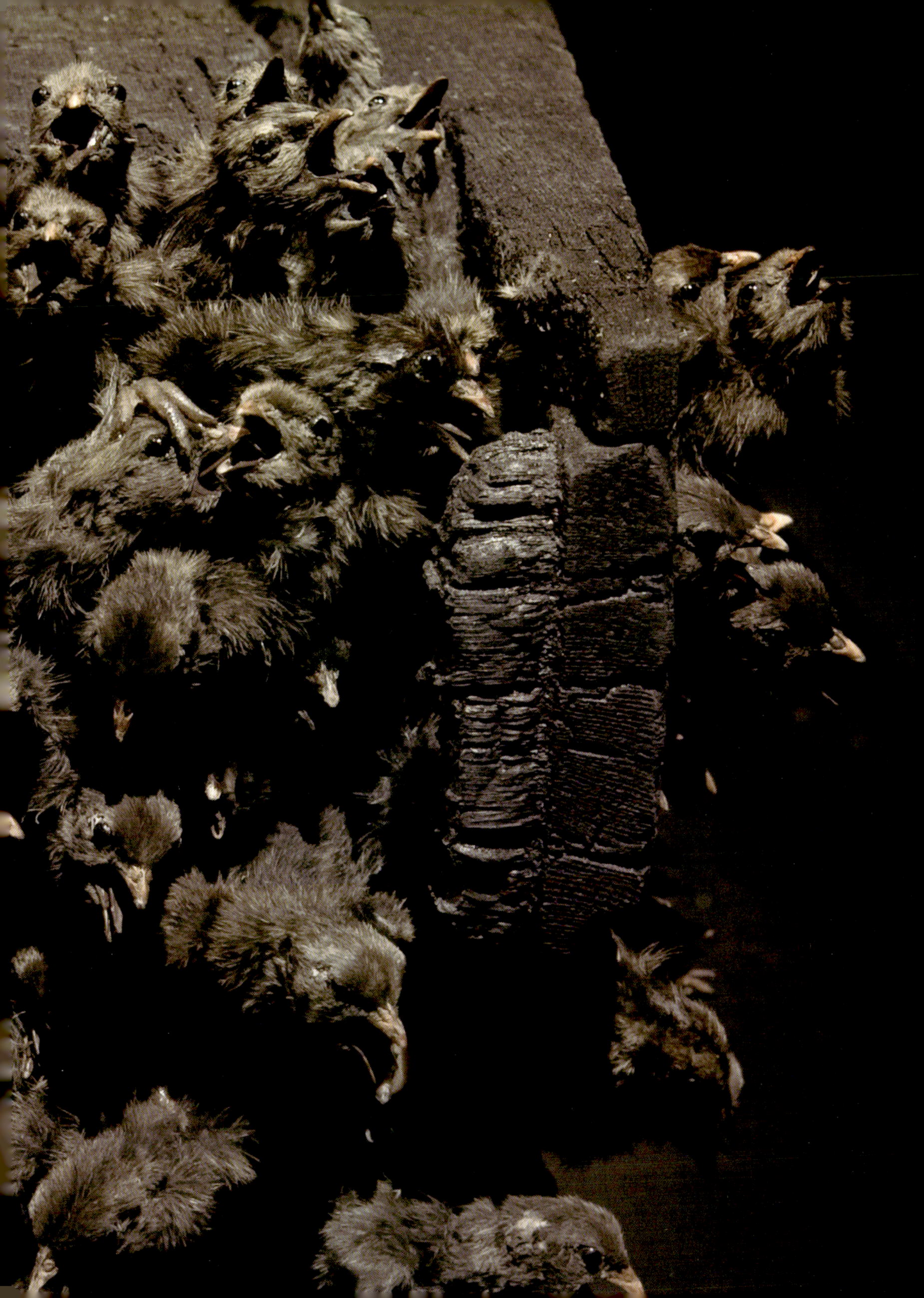

Dead Heads

= =

2009

- -

Photograph

- -

Dimensions variable

- -

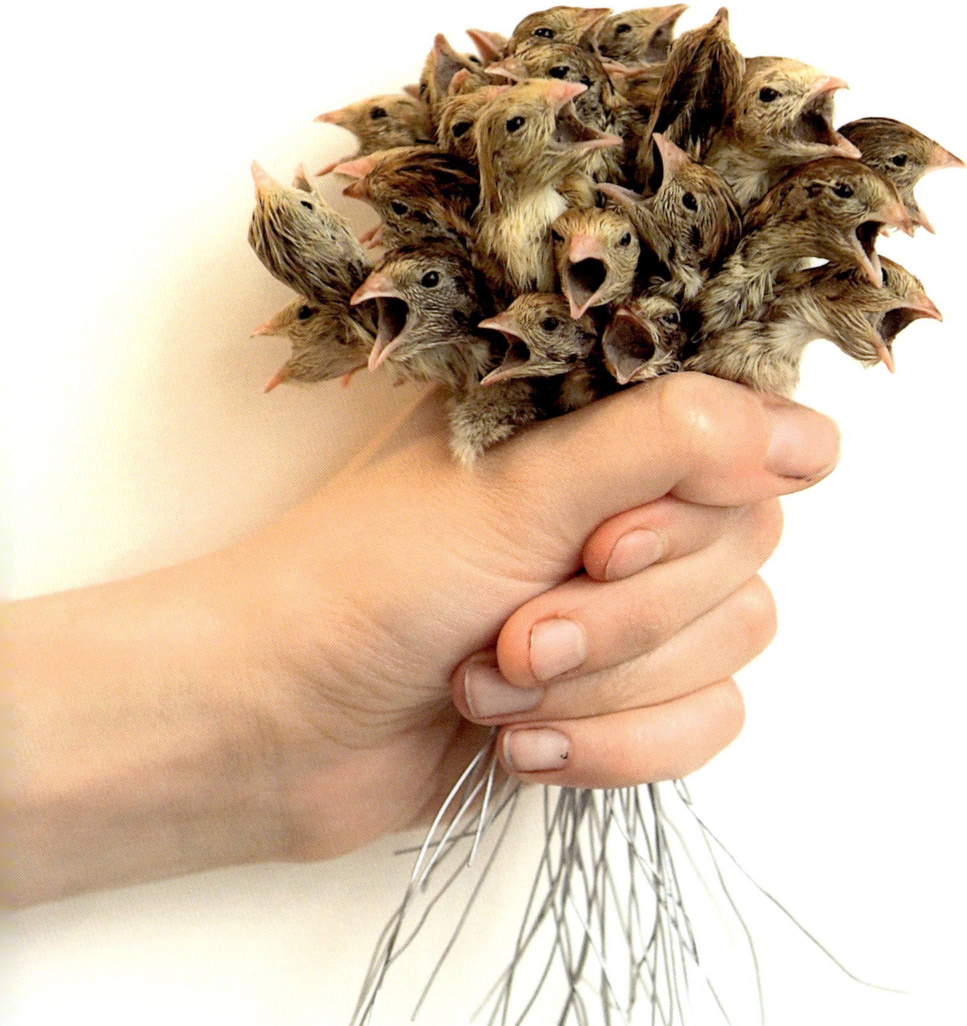

MS Found in a Bottle

= =

2008

- -

Taxidermy pigeon
Glass
Cork
Lacquered wood

- -

110 × 32 × 25 cm

- -

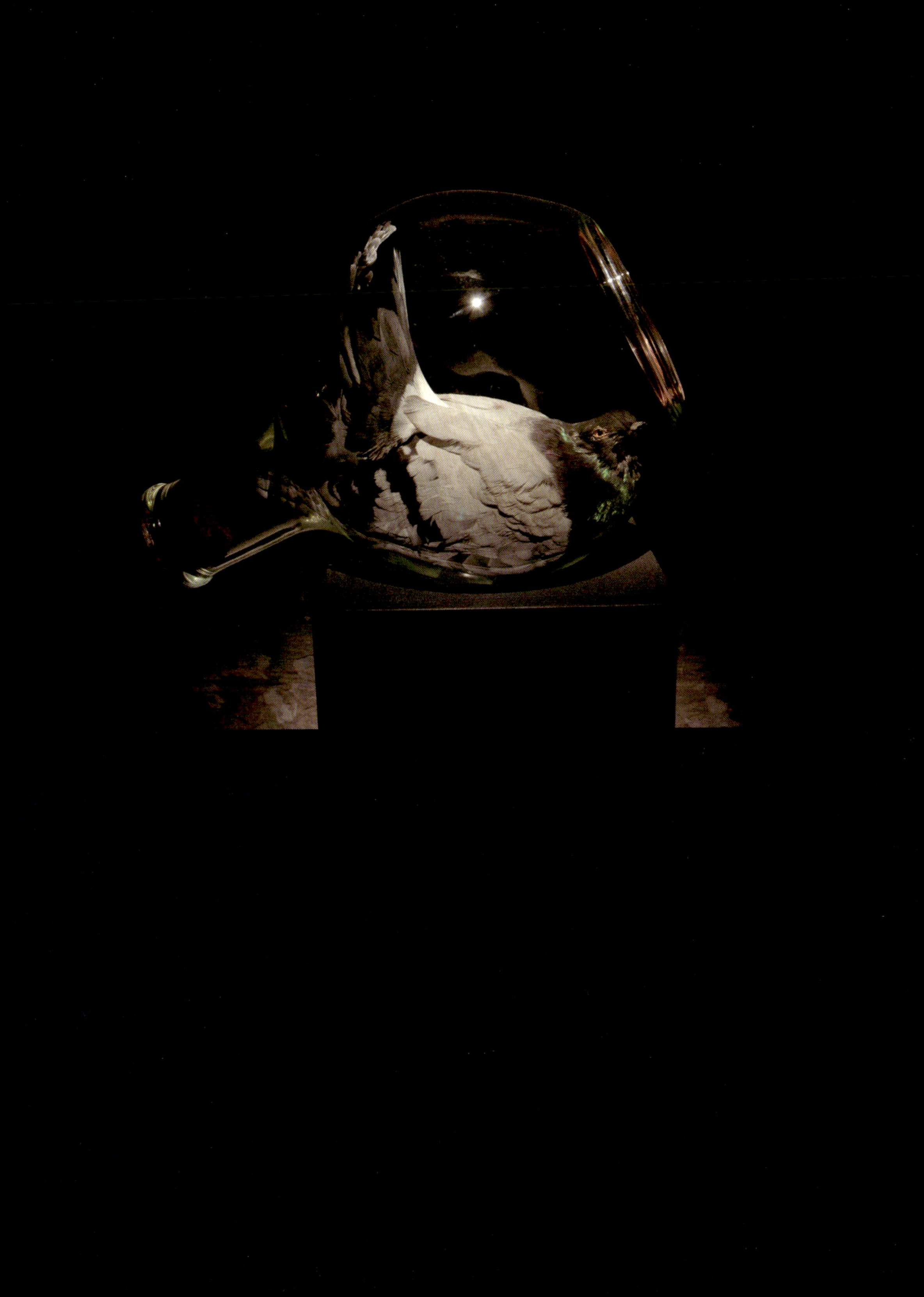

Someone on the Phone

= =

2007

- -

Taxidermy magpie
Bakelite telephone
Electrical wire

- -

32 × 17 × 20 cm

- -

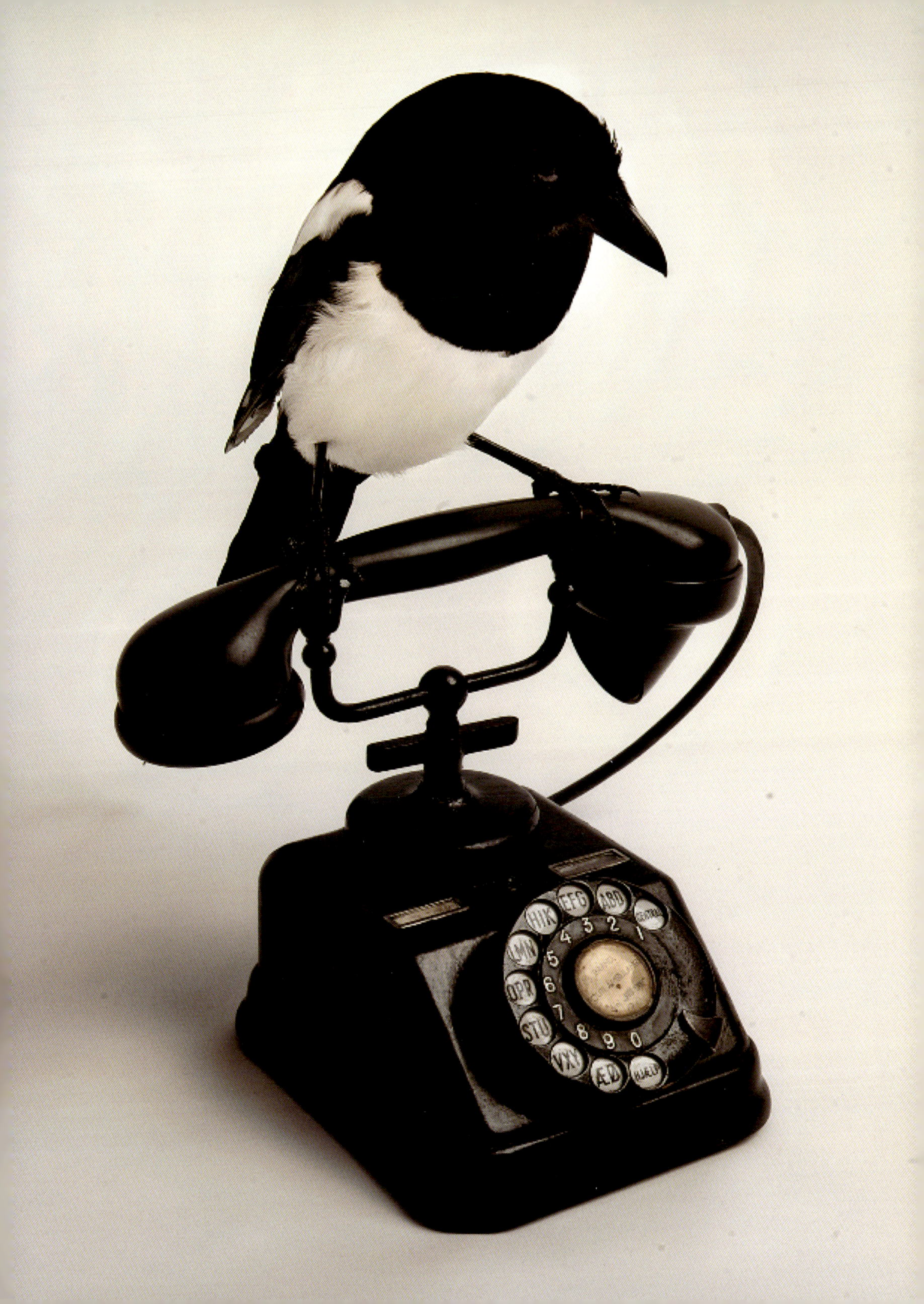

ABD
EFG
HIK
LMN
OPR
STU
VXY
ÆØ
1 2 3 4 5 6 7 8 9 0

Morning

= =

2007

- -

Taxidermy robin
Metal
Glass

- -

150 × 150 cm

- -

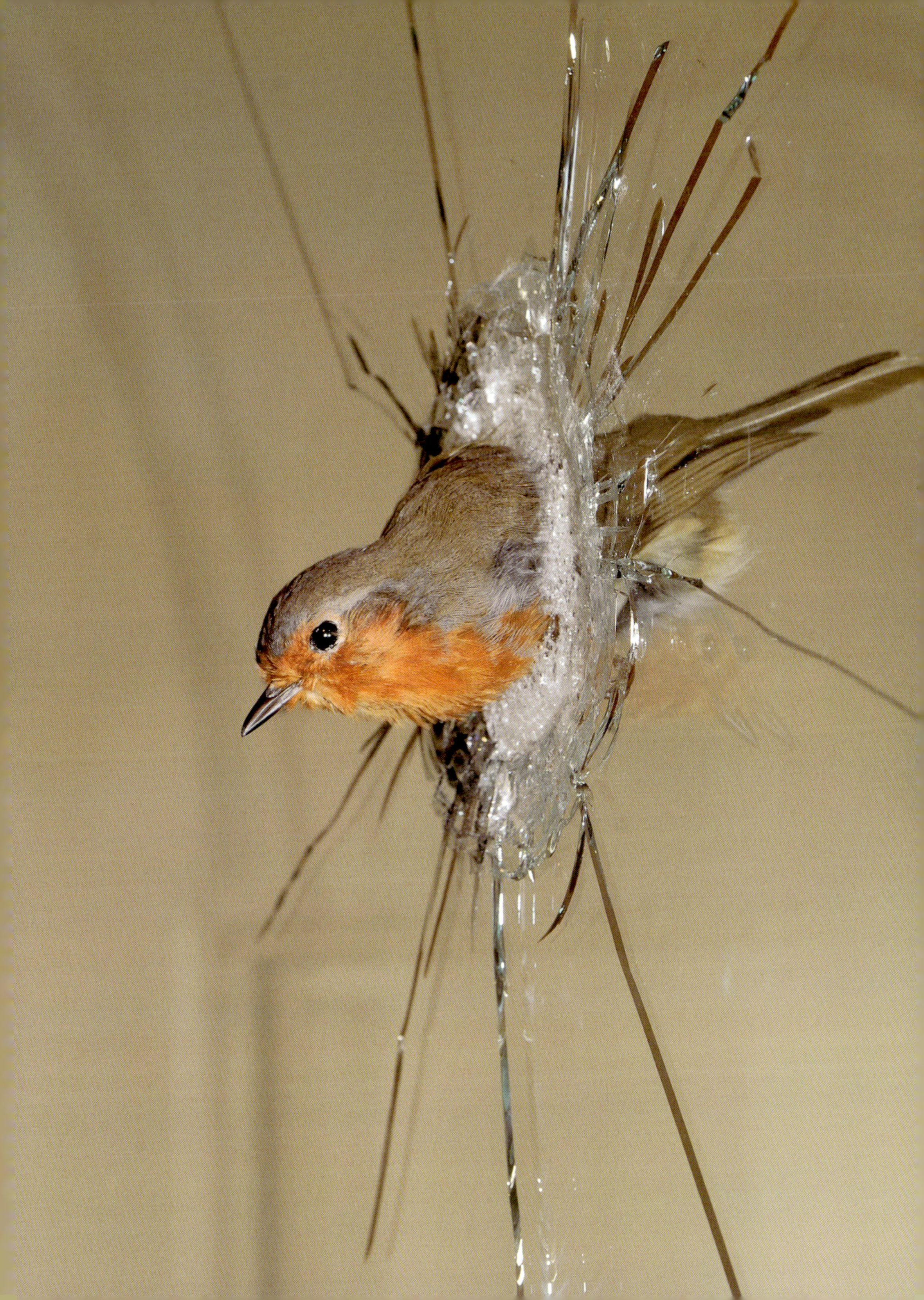

Former Things

= =

2007

- -

Taxidermy jackdaw
Taxidermy quail chicks
Glass panels
Steel
Wooden furniture
Cotton
Silver
Crystal chandelier

- -

184 × 184 × 184 cm

- -

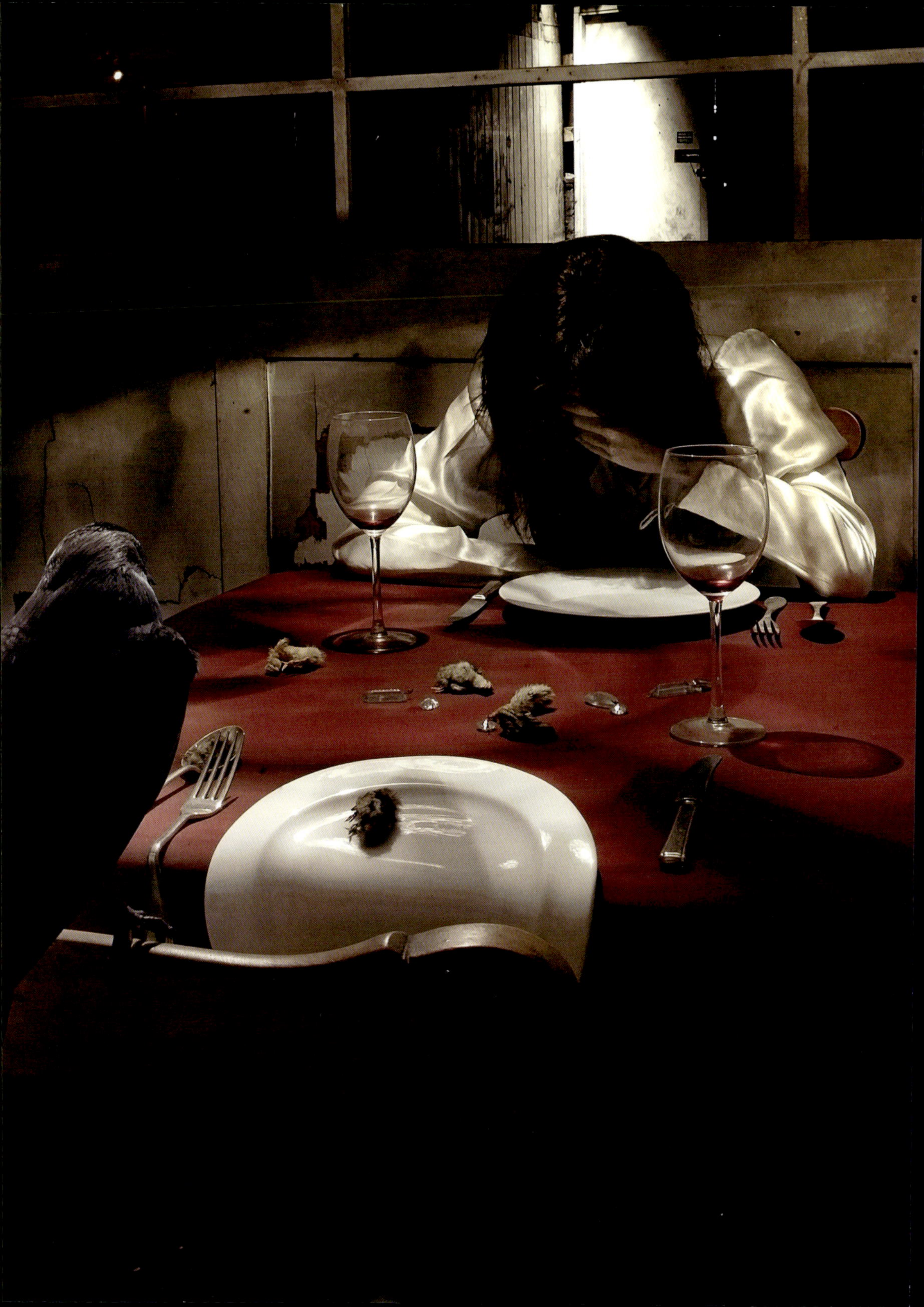

Testament

2007

Taxidermy robin
Crystal chandelier
Leather-bound prayer book
Glass
Wood

30 × 27 × 16 cm

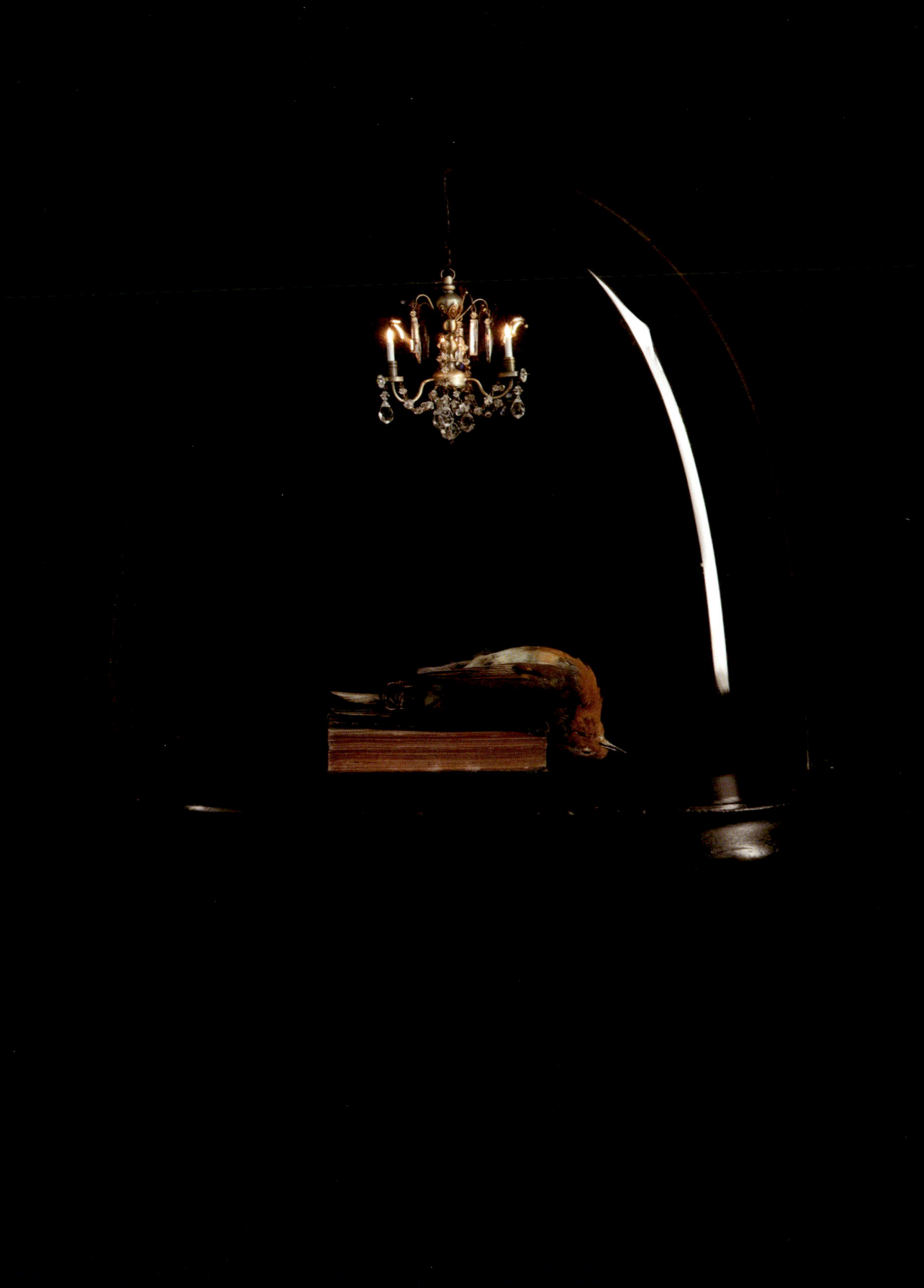

La Petite

2006

Taxidermy quail chick
Wood
Glass
Paper

20.7 × 15.2 × 8.1 cm

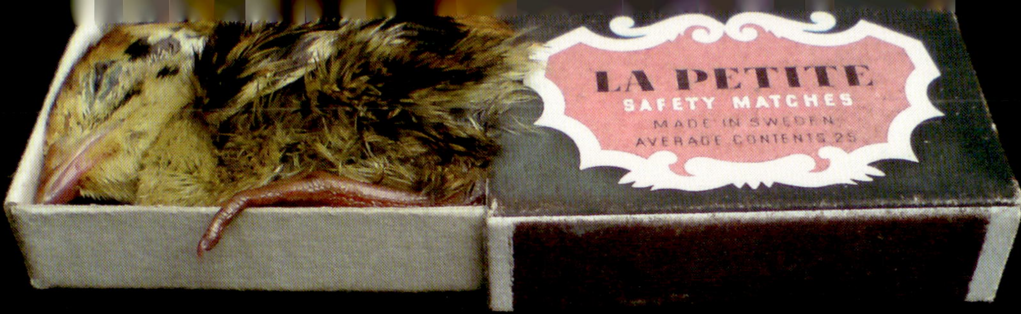
LA PETITE
SAFETY MATCHES
MADE IN SWEDEN
AVERAGE CONTENTS 25

Still Life After Death (Rabbit)

2006

Taxidermy rabbit
Top hat
Paint

158 × 28 × 23 cm

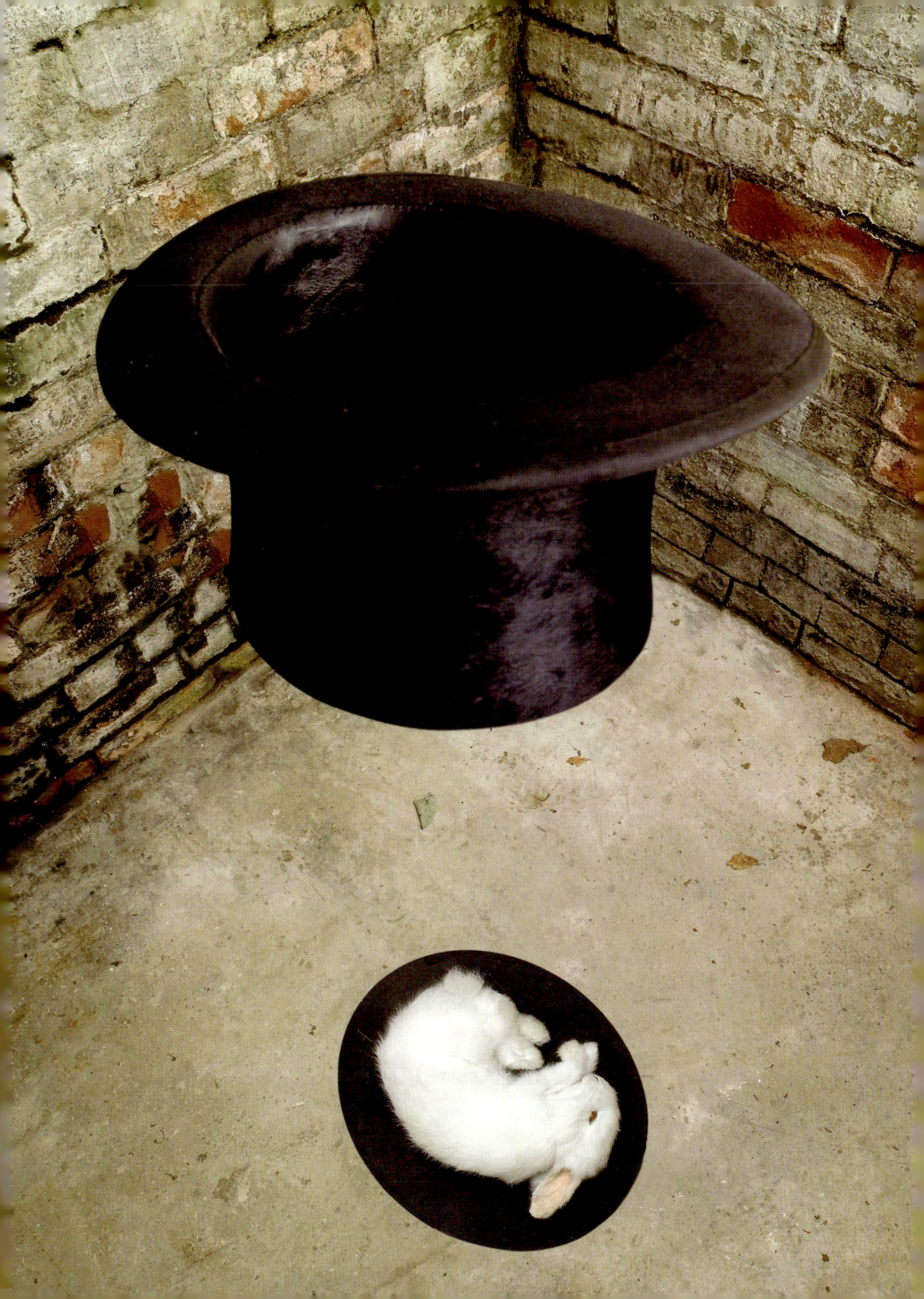

Still Life After Death (Fox)

2006

Taxidermy fox
Glass
Crystal chandelier

190 × 35 cm diameter

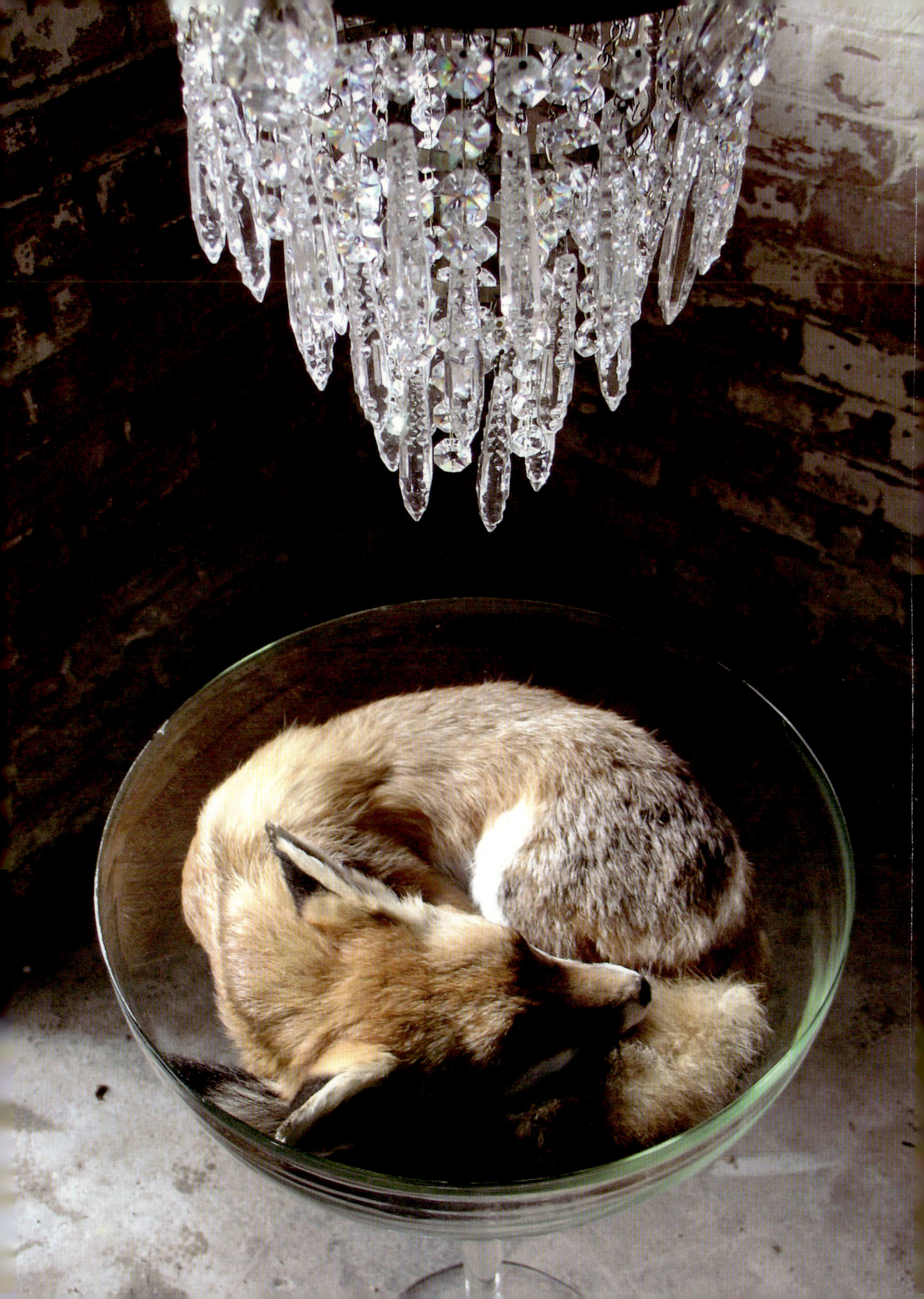

To Every Seed His Own Body

2006

Taxidermy blue tit
Crystal chandelier
Leather-bound prayer book
Glass
Wood

20 × 22 cm diameter

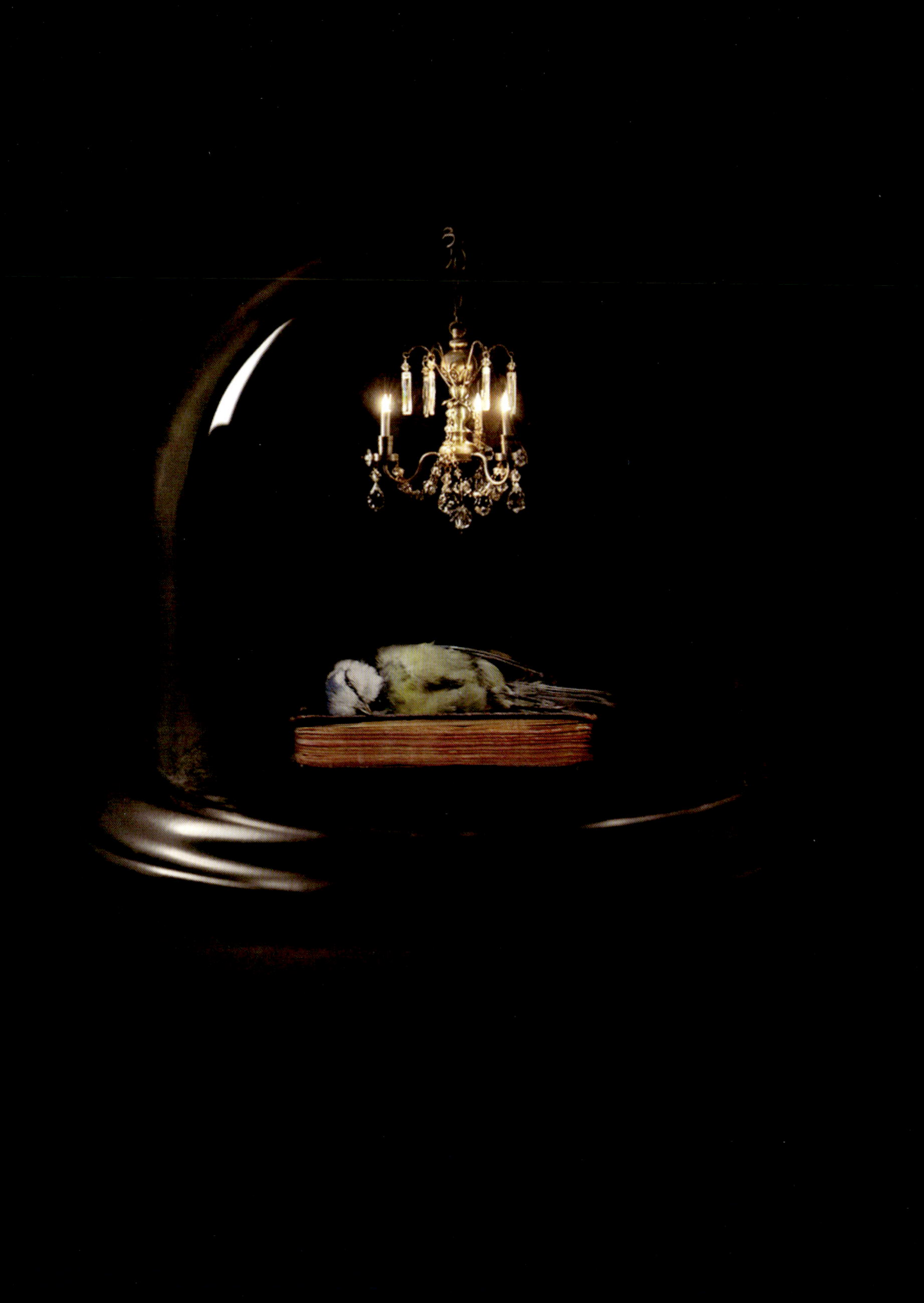

Rest a Little on the Lap of Life

= =

2005

- -

Taxidermy rat
Champagne glass
Crystal chandelier
Glass
Wood

- -

50 × 25 cm diameter

- -

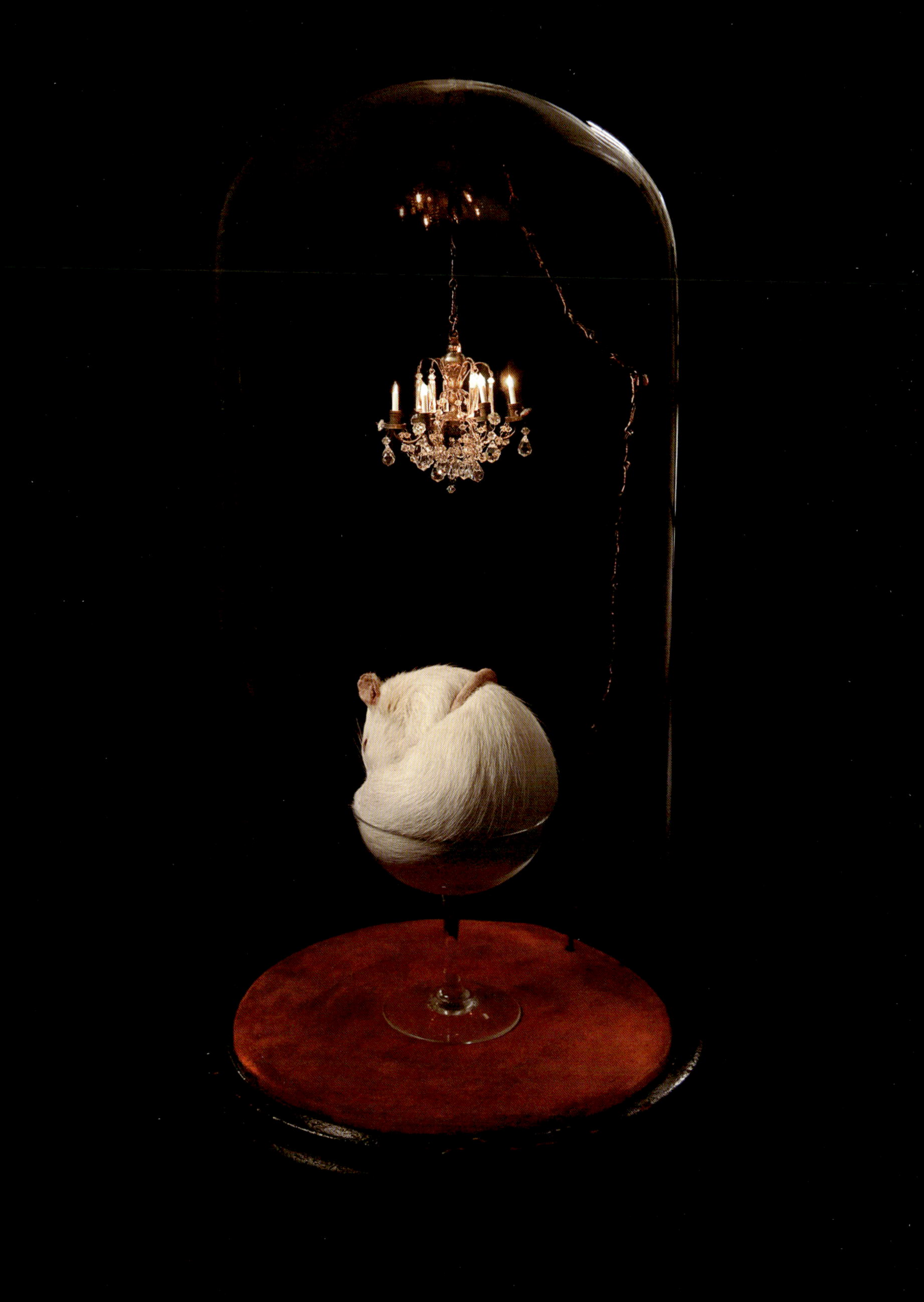

Lovebird

2005

Taxidermy lovebird
Taxidermy mouse skin
Brass
Metal
Glass
Wood

30 × 20 cm diameter

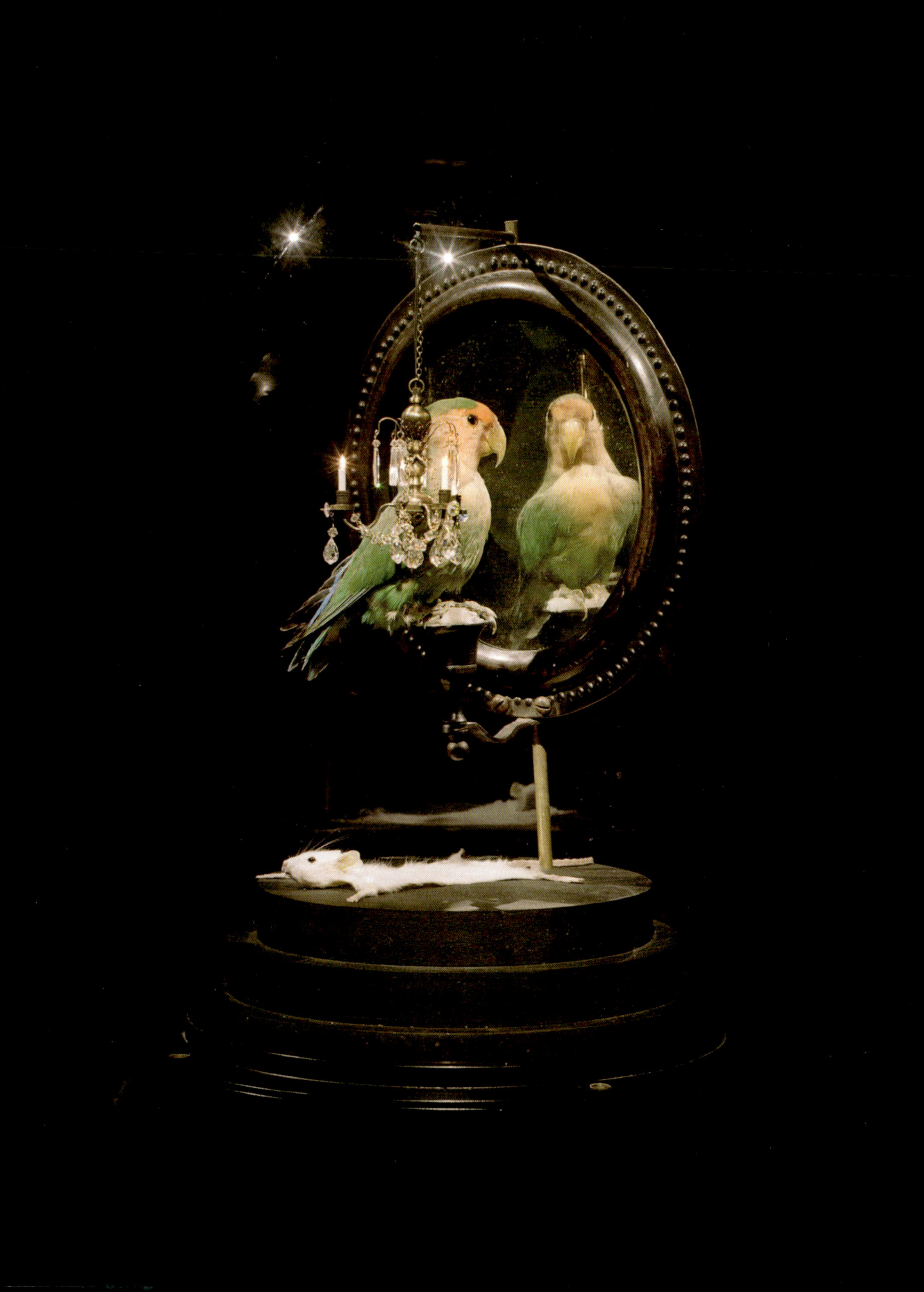

Polly Morgan in Conversation

with Guy Kennaway

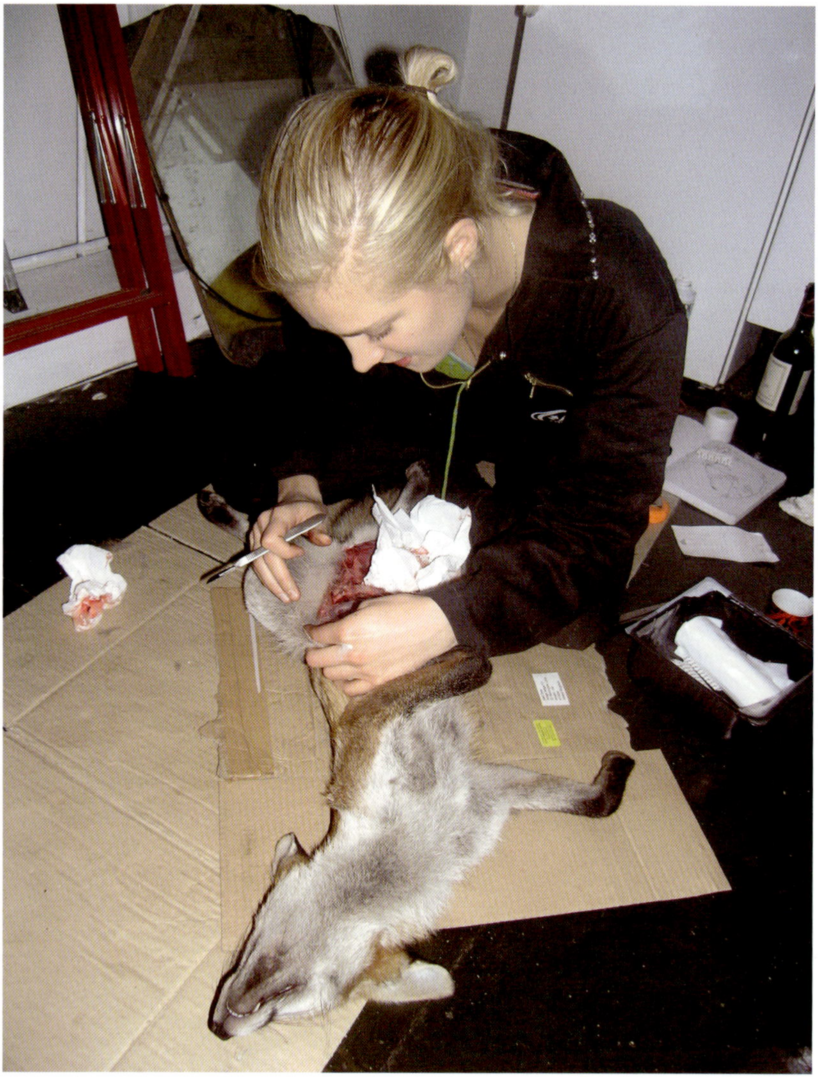

Guy Kennaway - Most of your work features taxidermy. What does an animal have to do to end up being in one of your pieces?

Polly Morgan -- Live and die. It can live as it likes, but it must die within sight of someone apt to post it to me (and preferably out of raging sunshine).

GK - Your work differs from conventional taxidermy, why is that? What are you doing that marks it out?

PM -- Conventional taxidermy mimics the natural habitat of the chosen animal; mine doesn't try to do that. I love traditional taxidermy but I've never been interested in working within those boundaries, probably because I'm competitive and have never wanted to be part of a crowd. If I had classified my work as taxidermy, alongside better taxidermists, I'd have been grouped together with people more adept than myself. I suppose having both old-fashioned and contemporary elements marks it out; the taxidermy and the belljars could be antique yet the cast balloons and the polished telephone receivers have a more Pop influence.

GK - When you use an animal, are you using its body, its spirit, or both?

PM -- Its body. Or, to be pedantic, its skin and coat. The actual body is discarded. I have nothing to do with its spirit. The animals in my freezer are my raw materials, like paint to a painter. There are millions of combinations to be made; the possibilities are endless. I don't dwell on the animal's life, I think to do so would be sentimental. I am interested in what it leaves behind.

GK - How useful has it been not to have gone to Art School?

PM -- I am happy working with taxidermy and that was less likely to have happened had I followed a different route. I think it worked for me as I didn't have a burning ambition to be an artist therefore I wasn't overawed by art-world people when I met them (in part as I rarely knew who they were), or encumbered by an in-depth knowledge of the history of art and the feelings of inferiority or unoriginality that can bring.

GK - Let's go back a bit; can you remember your first art experience? Were you aware of art as a child? What did you see, what did you like?

PM -- Until I was in my mid-teens and studied it at school, art to me was the little watercolours my Granny painted and the wire or stone sculptures I saw at local craft fairs. I was never happier than when making things but was quite solipsistic in the way I worked; I didn't have much interest in what others made, either in my class or in the wider world.

I was dimly aware of the Old Masters from history lessons and keen on what I knew of the Surrealists and Pop artists but hadn't a clue what was going on in contemporary art until I saw 'Sensation' at the Royal Academy when I was 17.

GK - What were the first things you made? Did you develop into an artist or did you arrive on earth as one?

PM -- I hate ornaments and never buy anything that I can't use, yet ironically have always been compelled to make things that serve no obvious purpose. When I was two I cut sheets of coloured paper into a pile resembling primitive confetti and insisted my mother helped me post it anonymously to a friend.

I have lain fallow at times in my life but I could never imagine not spending my time imaginatively. I often imply that I fell into art by accident but I'm probably being disingenuous; the fact I ended up an artist despite not going to art school and without pressure or encouragement to do so makes me think I may have been one already.

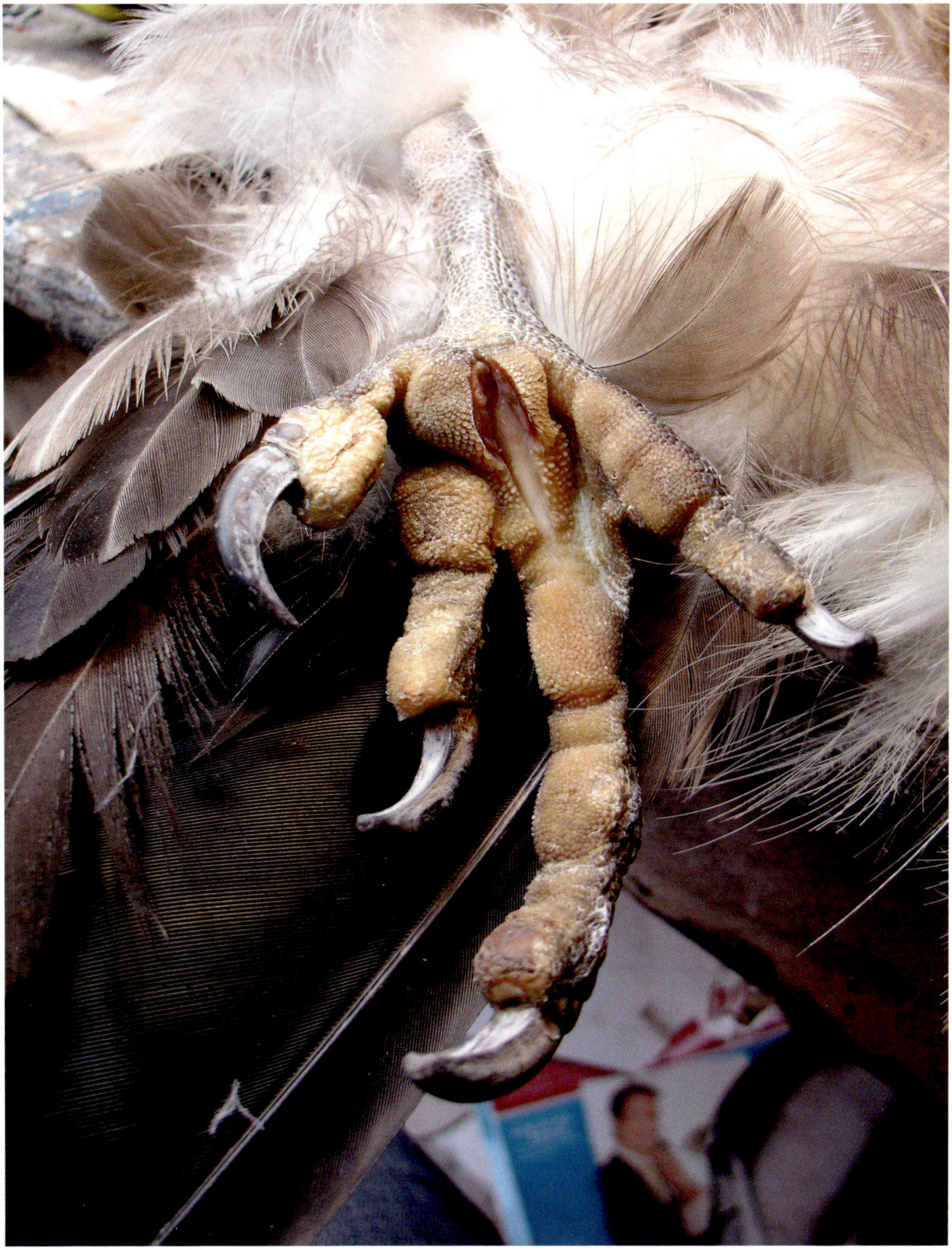

GK - You say you went to the 'Sensation' show – what was your experience? What did you like, or hate?

PM - - It was my first encounter with them; every name was new to me. I was 17 and on a day trip from London. The most contemporary artist I'd had introduced by my art teacher at school was Andy Goldsworthy so I was shocked and delighted when I saw 'Sensation'; everything there was new to me and I couldn't believe it was ok to show children with cocks for noses in a gallery. I'd always loved irreverence but it was rarely allowed and I'd trained myself to stifle it. It was exciting to discover a place for it.

I loved Ron Mueck's *Dead Dad* just for its smallness and detail. I can't necessarily explain why but when I see a huge painting I like I always think 'I'd love to have that in miniature', I find it hard not to fantasise about shrinking my own larger works.

GK - Let's talk about the flying machine, which is a recent, ultra confident work, a kind of coup de theatre, with a lot of narrative in it. I saw it in the church in Marylebone Road during Frieze, where it dominated the room. It has got a definite story in it, and I often find that the things you do have a narrative, like short stories almost, and I have puzzled long and hard over the chicks emerging from the coffin in *Carrion Call.* Is there a story in your head, or just mine?

PM - - Not a story exactly. *Carrion Call* was inspired by a photograph I took for the ICA's 60th anniversary book (themed 'Tomorrow') of a rotten blackbird riddled with maggots. I wanted to avoid clichés of 'tomorrow' so I chose to photograph something repellent that was, on consideration, actually life-affirming and oddly beautiful; a corpse becoming a nest for new life. A coffin is loosely egg-shaped anyway, which is an amusing thought and I wanted it to look like it was cracking open with an overwhelming number of chicks all clamouring to emerge. They also looked like a growth, moss or barnacles from a distance – another sign of life from dereliction.

- -

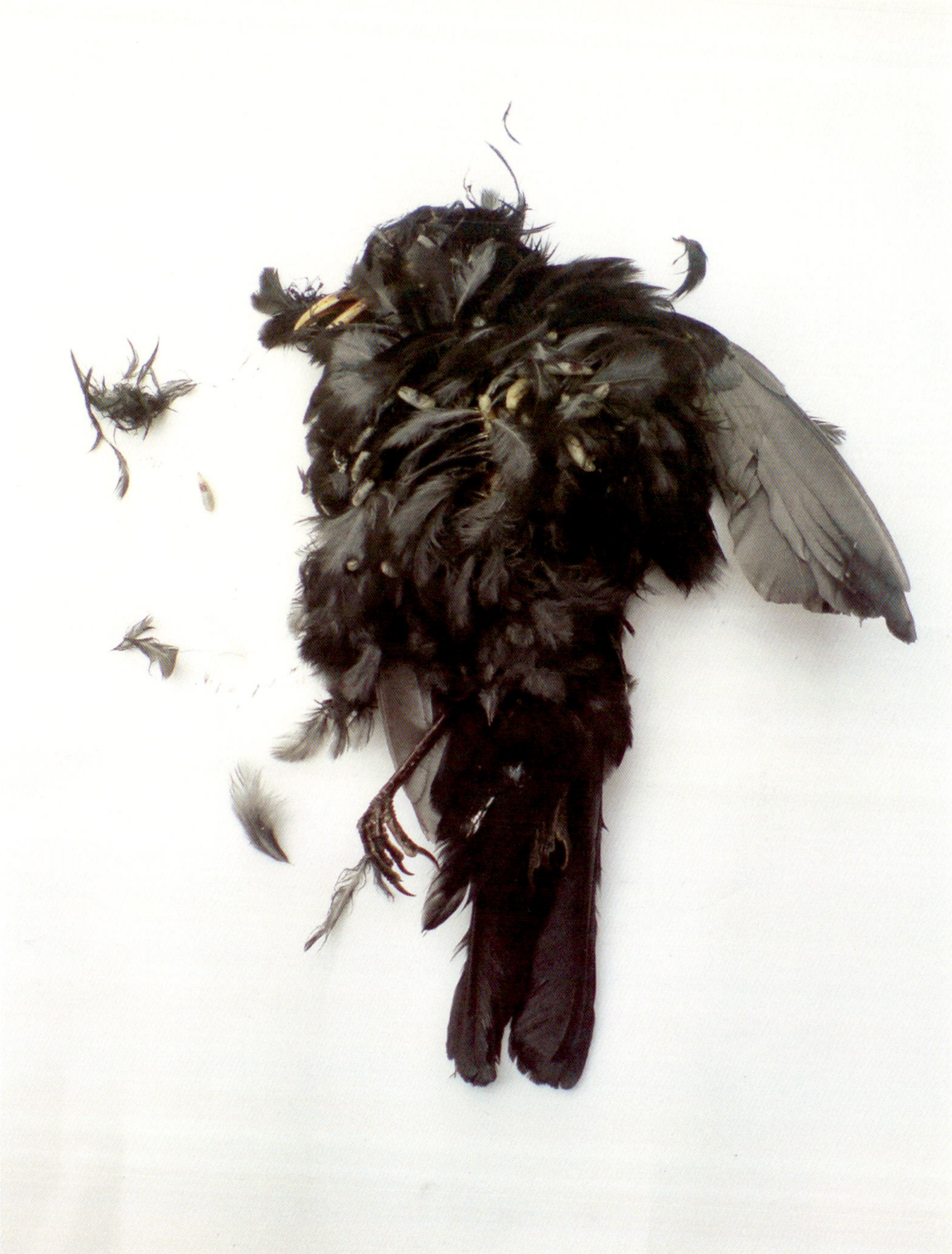

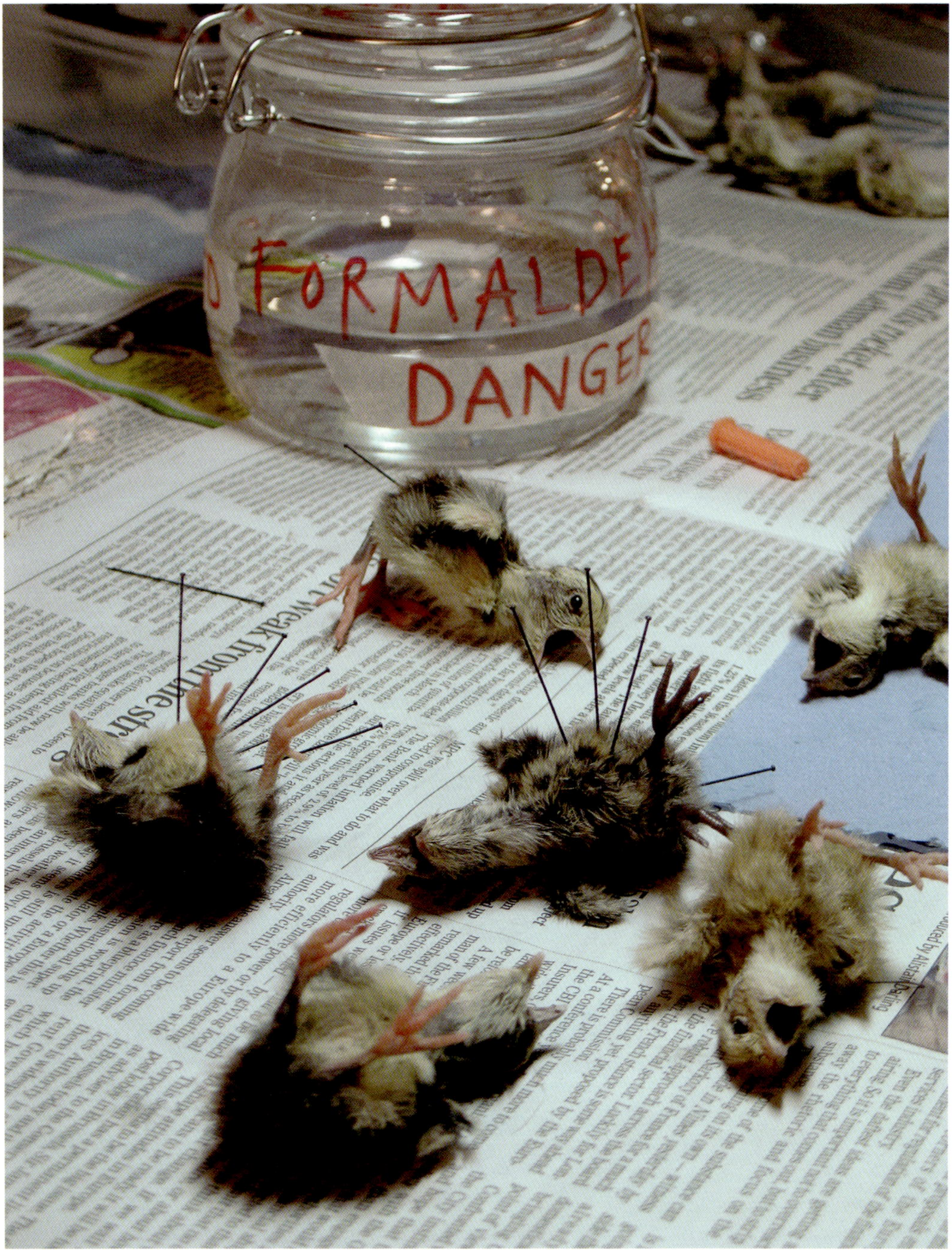
FORMALDE
DANGER

I also saw a programme on TV called 'Swarm' about creatures that congregate en masse: starlings, shoals of fish, butterflies and bugs, etc. I noticed that anything, however benign, could look terrifying if you multiply it enough – it becomes an army, and instinct tells you you're under attack.

GK - I have next to me a piece of yours which is a yellow tit lying on a cushion under glass. I was at a funeral the other day and I thought of it, because you are celebrating its life, by honouring it in its death, and I start to think about how it was, how it flew, how it lived and how it died. Does that seem a rational response to it? Or is it less about the animal as an individual soul?

PM -- I think that seems rational. I confess I think less about their souls the more I handle them. I prefer live animals but I have this selfish overbearing urge to hold them and examine them and just generally do all the things they hate. Dead, I can spend as long as I like looking – I just love the fact they're not going to rot or fly away.

GK - I often think your studio is like a science lab, with you as the biology teacher and your assistants as hard working students; it has the chemical smells of a lab rather than a studio, and all these specimens lying around in plastic bags. I notice that you recently got yourself a dog – a live one that is. I saw him sitting on your sofa watching all the work and I wondered if he was thinking, 'I wonder when it's my turn to be stuffed and placed under glass?'. Does your work make him nervous?

PM -- He's pretty impervious to food generally so doesn't have much interest. He has the odd sniff. I was hoping he'd be my very own canine disposal unit and I keep trying to feed him the bodies I remove from the birds but so far he's not much taken with them. I've been told it's the healthiest diet he could have.

There is a definite science lab element to my studio, I love that side of it as I always coveted the paraphernalia of science but never had

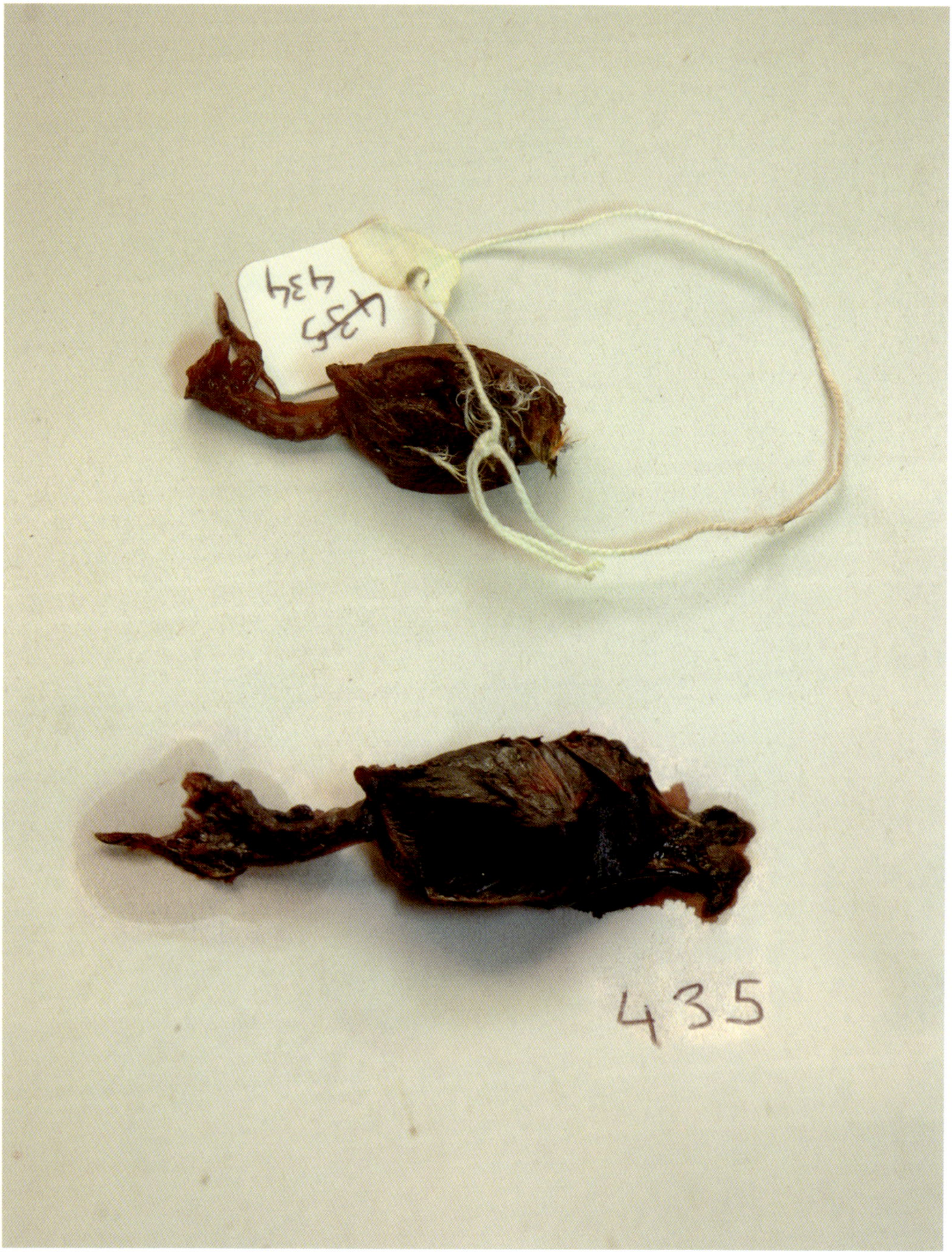
435
434
435

any use for it. I usually have just one assistant but in the run up to a show I take on several interns, all of whom are keen to learn so I do feel like a teacher some of the time, a role I thought I'd never see myself in!

GK - You are very good at it, and I can tell they respect and love you. I saw in your studio some birds attached to single balloons under glass. This struck me as a new theme: the pieces looked more formalist, if that's the right word, they had a strong shape rather than a strong emotion. What inspired them, and are you going to develop this theme?

PM - - I think it's natural to grow tired of your own work and push it in different directions and I've got a bit sick of the Victorian aesthetic. It's not that I don't like it – it's apparent from most of my work that I do, it's more that there is a lot of it out there now and I want to modernise taxidermy and combine it with much more contemporary elements. The balloon and chick series are simple but I think the combination of textures/shapes and colours of the chick versus the balloon work really well. Both are so tactile and uncanny; the balloon will never deflate (it is cast in resin) and the chick will never decompose. For me the chick represents the Victorian aesthetic, the balloon a more modern, Pop aesthetic. The title, *Still Birth*, is not designed to shock but it seemed so apt considering the work is a still life with a dead chick, and it's attached by a cord to a womb-like red balloon.

My work for the 'Psychopomps' exhibition, *Atrial Flutter*, takes this further, by representing death in the form of a red cardinal huddled, where the heart would have been, inside a human ribcage. Balloons are tied to the ribs and carry it off.

GK - You've also made flying machines. What was the inspiration for these?

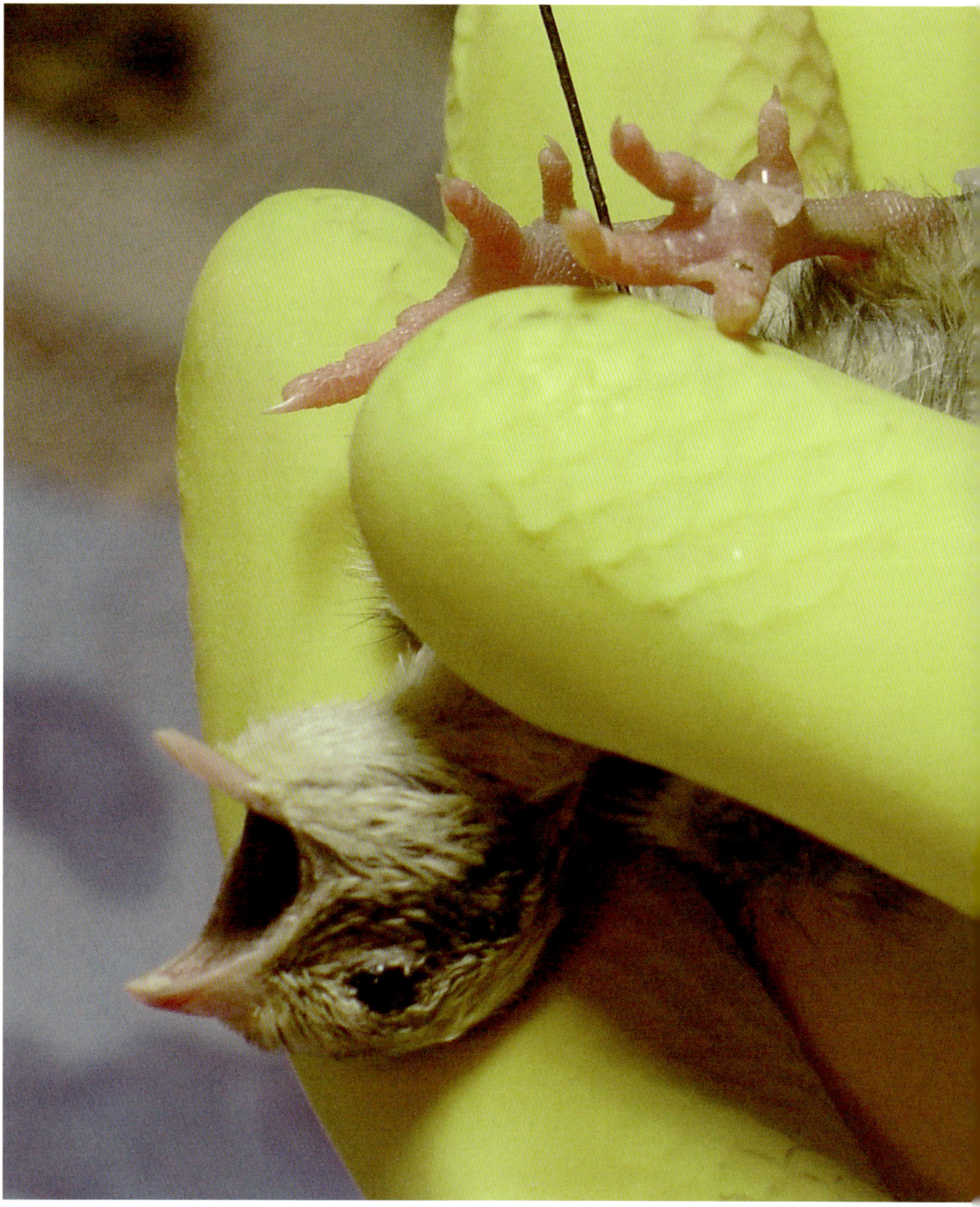

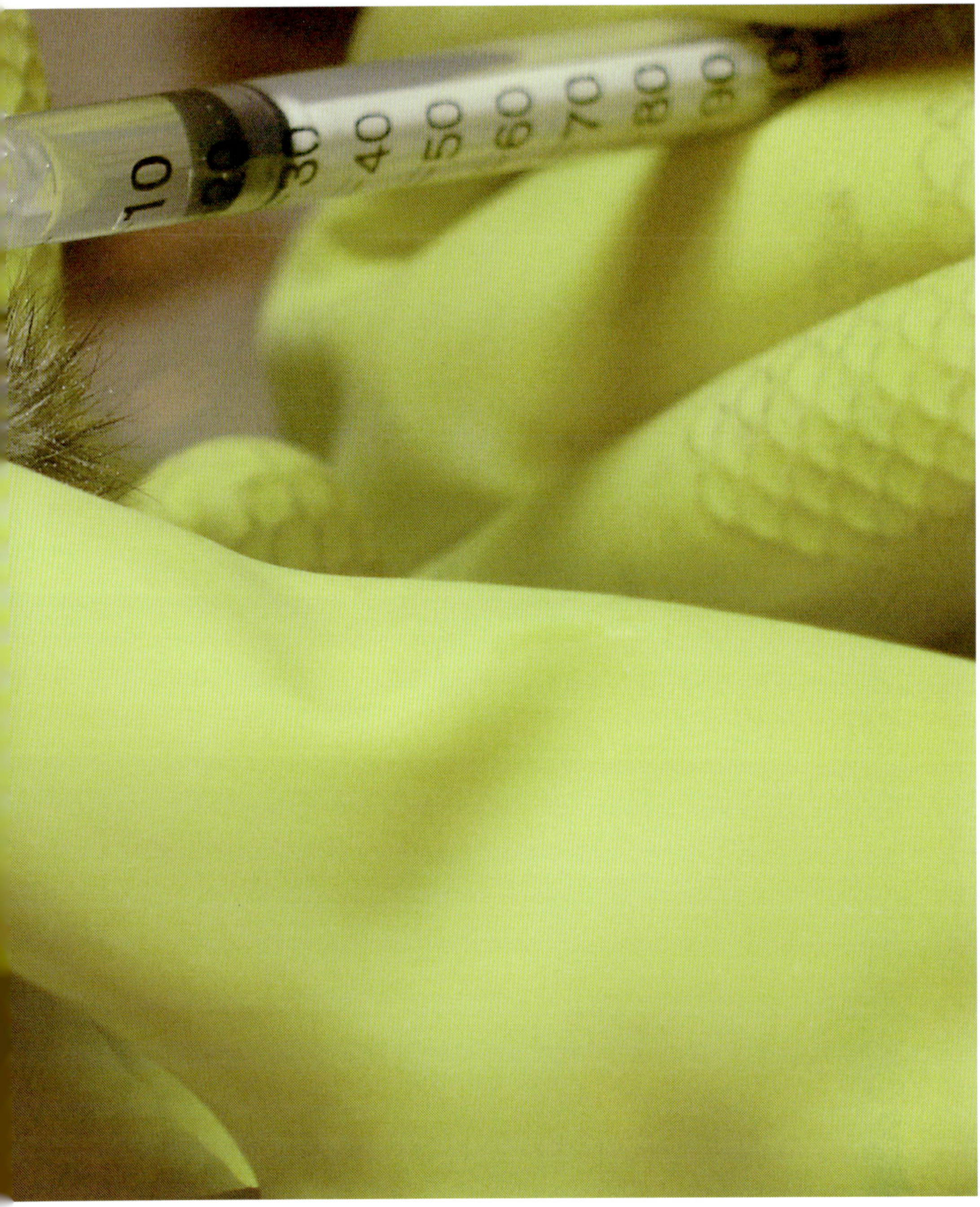

PM -- I read a book called 'Flying's Strangest Moments' by John Harding, which comprehensively covers all recorded attempts by man to fly. I was touched by the men who'd fashioned primitive flying machines out of strap-on wings and balloon-clusters tied to their chairs in their obsessive pursuit of flight. I think that from the day we become conscious of space and the never-ending sky above we feel either defeated or inspired. It is people who fall into the latter category, and I'd count the inventor of the eagle-led carriage on which I based my first flying machine *Departures* among them, I am interested in.

The second flying machine I have made for 'Psychopomps', *Systemic Inflammation*, is made to look like it's ascending from the fires of hell. I wanted the birds to look like flames, only the colouring of those I had wasn't the right shade. We experimented with countless brands of ladies' hair dye, even combining them to try to find the correct hue. For a while it was like a bird's beauty salon in the studio; some being dyed, some washed and others blow dried.

Getting the finish of the metal cage right was equally trying. It was first rusted, then painted black, then burnt with a blowtorch. The paint didn't bubble and flake as dramatically as I'd hoped so we tried putting various tapes onto the metal: electrical, packing and gaffa, and burning them off. This started to look better but still wasn't textured enough. We were about to give up when I saw some masking tape lying around and thought it'd be worth giving that a go. It blackened and peeled in just the right way and the whole afternoon was then spent mummifying the cage in masking tape; making it look pale and pristine before blasting it all off again.

GK - Tell me about the two more abstract, winged works you've made for 'Psychopomps'?

PM -- They're called *Blue Fever* and *Black Fever* and are an amalgamation of the balloons and birds in flight I had adopted in previous work. I wanted to make an organic whole that looked uncannily real. I looked at Eadweard Muybridge's time-lapse images of birds in flight and decided to make the different strata of wings in my sculpture mimic

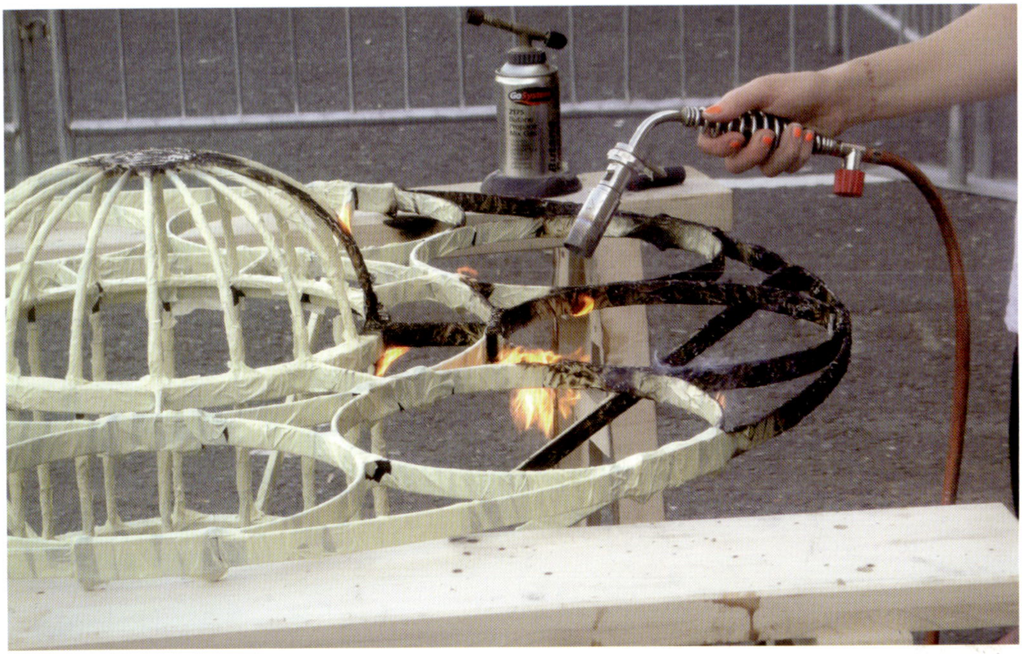

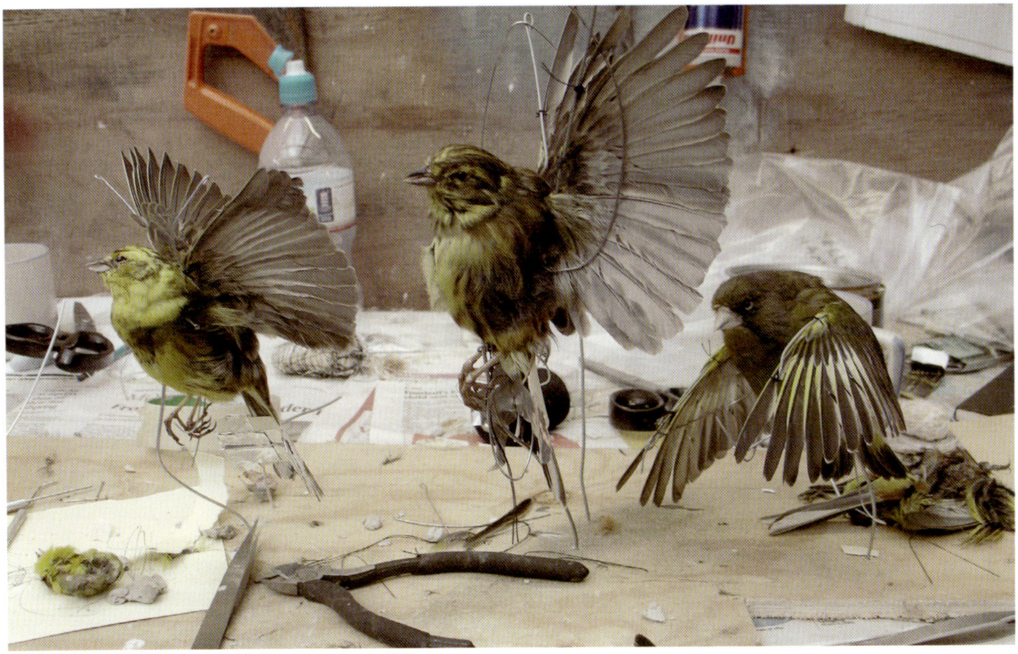

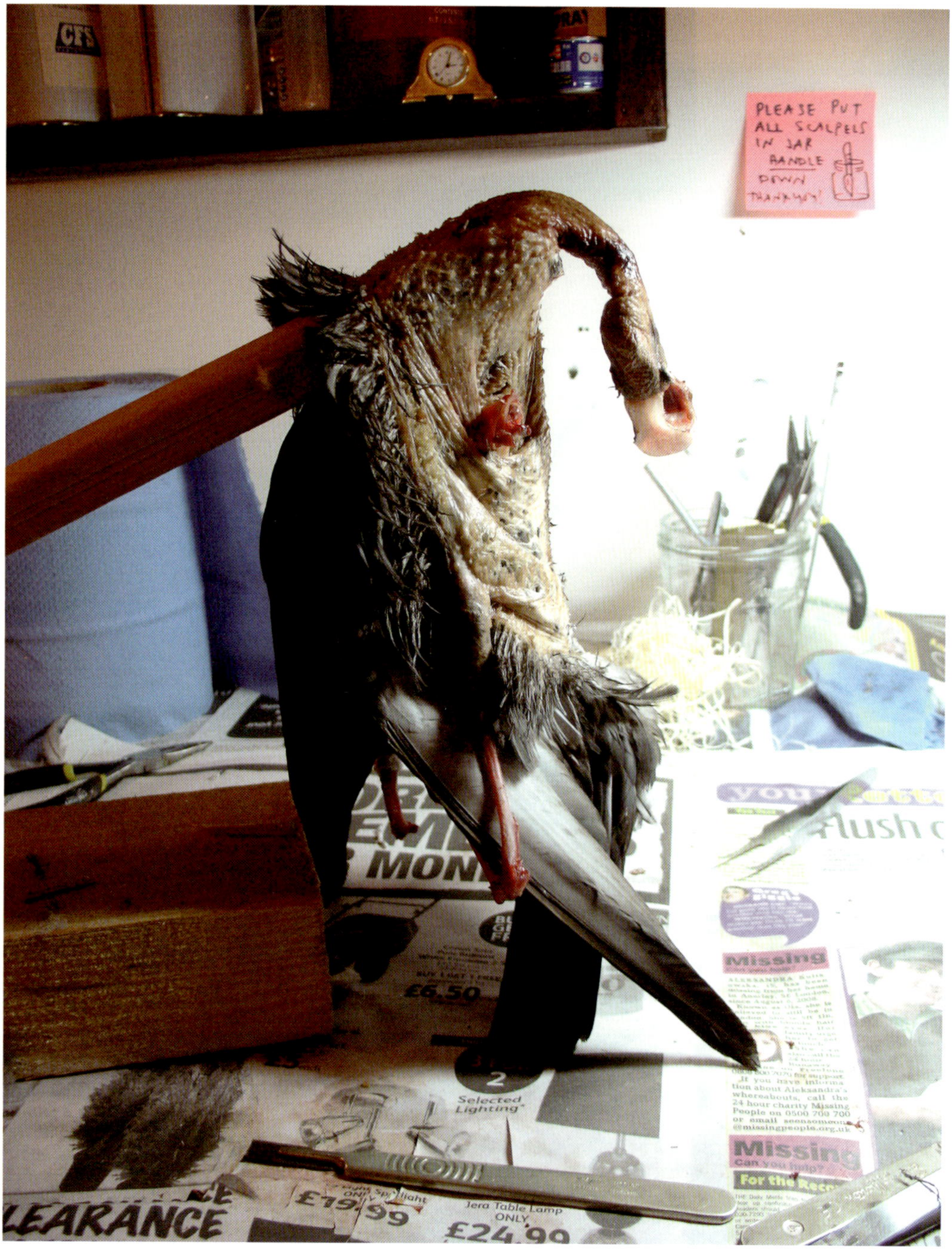
PLEASE PUT ALL SCALPELS IN JAR HANDLE DOWN THANKYOU!
Missing
Selected Lighting
£6.50
Missing
Jera Table Lamp ONLY
£24.99
LEARANCE

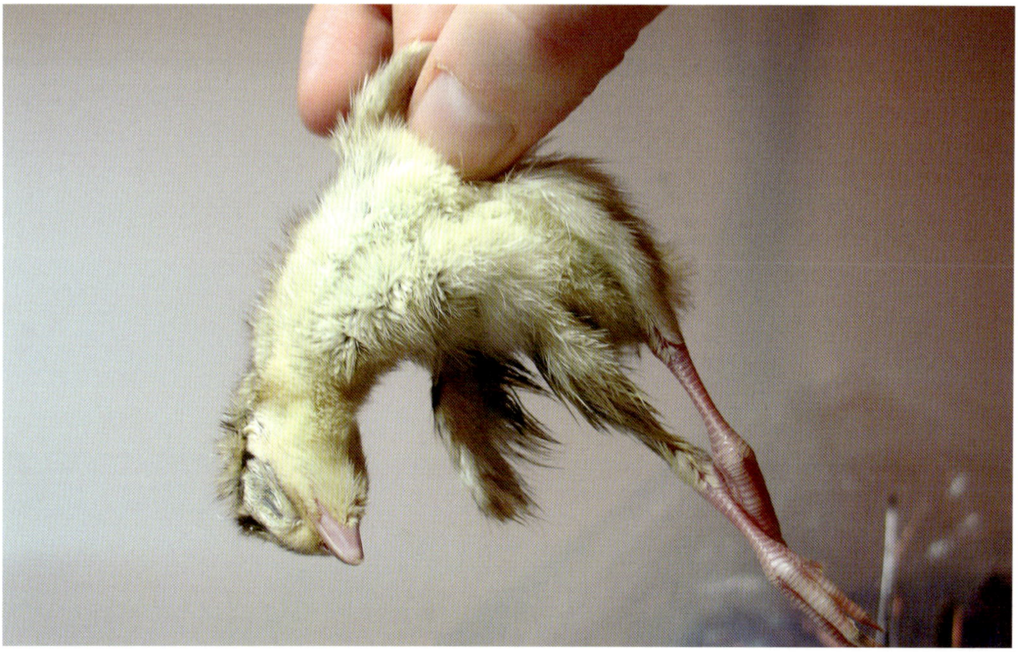

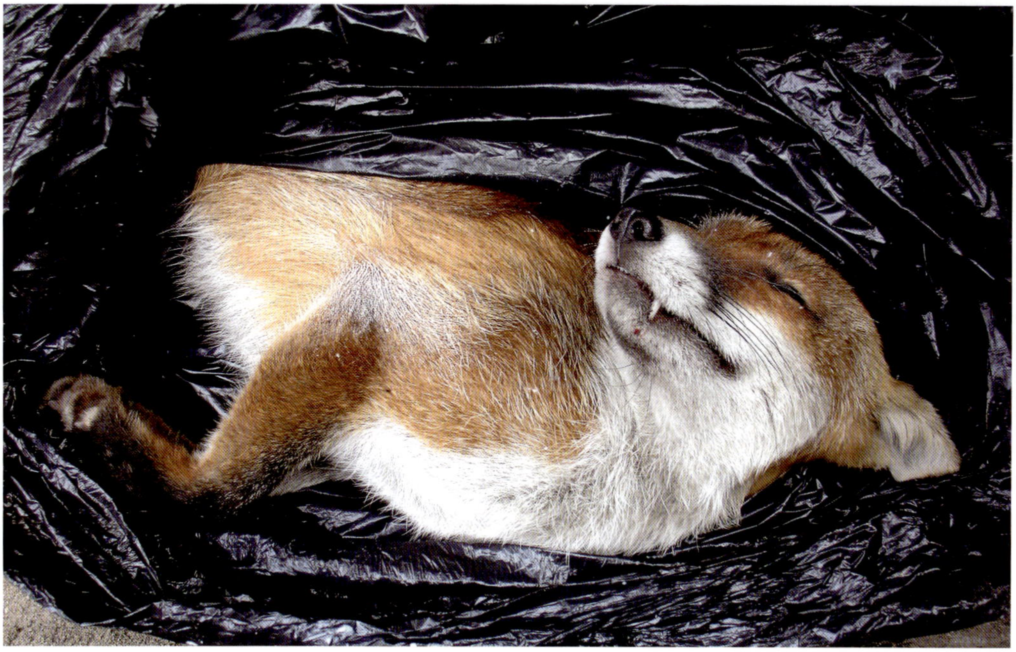

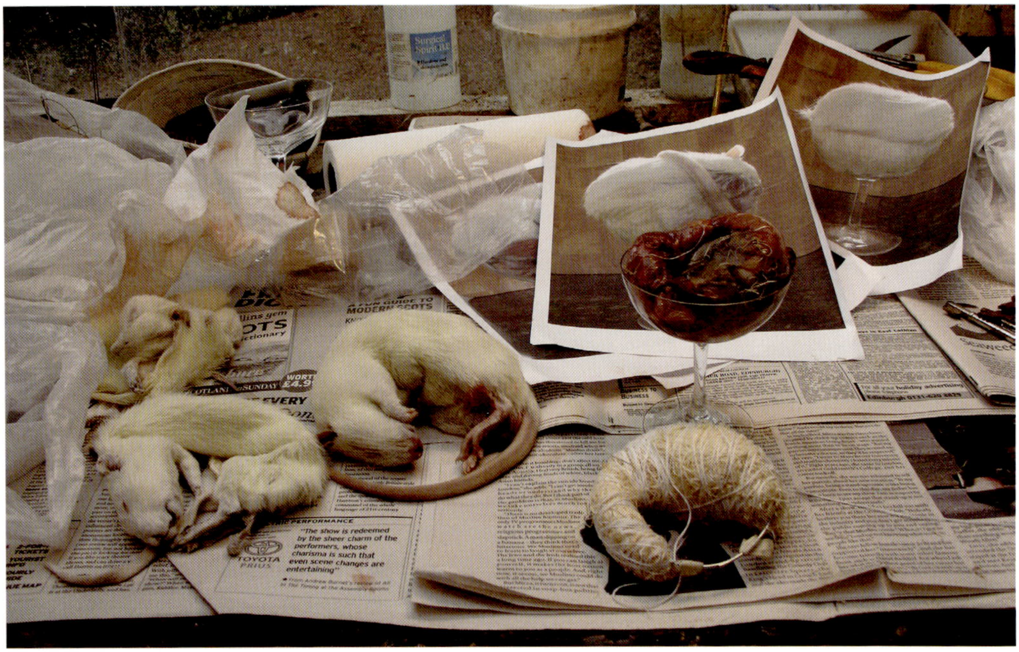
PERFORMANCE
"The show is redeemed by the sheer charm of the performers, whose charisma is such that even scene changes are entertaining"
TOYOTA
PRIUS

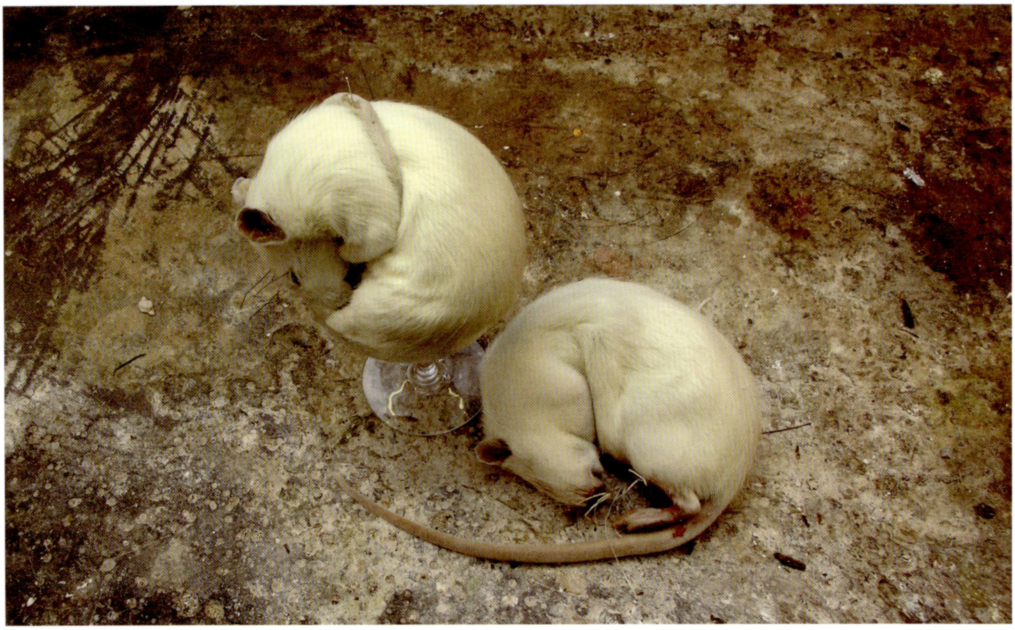

these stages, so that the objects seemed to have a pulse. We skinned 60 crows and 60 pigeons and removed the right wing only from each bird. We prepared them as though for traditional taxidermy, with hemp binding a wire to the bone and bulking out where the flesh had been. A frame was built to fix the wings to, with each entry point marked around each ring. The frame was filled with woodwool and bound with twine before each skin was stitched to the adjacent one.

GK - In your earlier work the animals are really at peace, whereas the birds in 'Psychopomps' and in a work like *Morning*, which show a robin smashing through glass, are more active.

PM - - I think I just moved on, I couldn't go on making the same point forever. There is so much you can represent with taxidermy and I wanted to explore other possibilities. In *Morning*, I saw the glass as representing the end of life and start of death.

GK - I was also struck by the amount of glass in your work. Aside from practical considerations, in terms of displaying the taxidermy, I felt there must be a reason why you chose to use belljars, glasses, bottles, etc.

PM - - I've always loved glasses. I used to go to charity shops and buy odd glasses that took my fancy. They're like transparent moulds, restricting and sculpting the form of whatever is inside them – i.e. a glass of water becomes water in the shape of a glass. In *Rest a Little on the Lap of Life*, I loved the way the rat's form changed as it entered the glass from having its natural rolls of flesh to forming a perfect hemisphere. There's something grotesque and compelling about the flesh of dead mammals and how they take on the form of whatever you put it on/in.

GK - Most artists tend to thrive in groups or movements. You seem to me to be one who ploughs her own furrow.

Do you want to be in a group? Are there artists who you consider to be in the same movement as you? Which ones would they be?

PM -- Paul Fryer and Wolfe Lenkiewicz were influential at the inception of my art life and we have continued to walk similar paths and be good friends, but I'd hesitate to call us a movement. There are other artists I have shown alongside regularly but outside of that we don't all socialise much. Doesn't history decide whether you were part of a movement or not?

GK - It often seems to me that art movements have become artificial marketing constructs.

PM -- I think that since the YBAs they have been. I was asked by the Evening Standard to appear in an article about the art 'group' in East London. They wanted to photograph me with other local artists and no doubt give us some unimaginative moniker for the purposes of the article only. I'd only ever met one of the artists involved and it was all so staged I turned it down. I'm relieved not to be part of a group or 'movement', at least not wittingly.

GK - You can see how you have evolved into new directions as you gained experience.

PM -- It sounds pretentious but I think art is like another language and when you start out you're not fluent and you struggle to be understood or get the cultural references. The longer you inhabit the world, the more able you are to communicate and the wider your vocabulary becomes.

GK - Was it like having two parallel progressions for you – as a taxidermist (your increasing confidence and skill with the craft) and as an artist (applying that craft)?

PM -- Very much so. I wasn't confident with the taxidermy at the start and as I've got better I've allowed myself to become more ambitious as

I'm less likely to let myself down. It's perhaps been a slower progression than for an artist who outsources a lot of the work but it's allowed me more time to mull over my ideas and weed out the bad ones.

GK - I read somewhere that you weren't happy with being called a 'taxidermist' as you felt you didn't meet a certain standard – would you say that has changed now?

PM - - It's not only the standard thing but it also gives the wrong impression; that I'm mounting peoples' pets and hunting trophies and am available for commission. Another reason is that I don't have the same priorities as taxidermists, who want as precise a representation of the animals as can be... I sometimes manipulate their forms to my own ends – a taxidermist would never do that. So I prefer sculpture, it's a broader term.

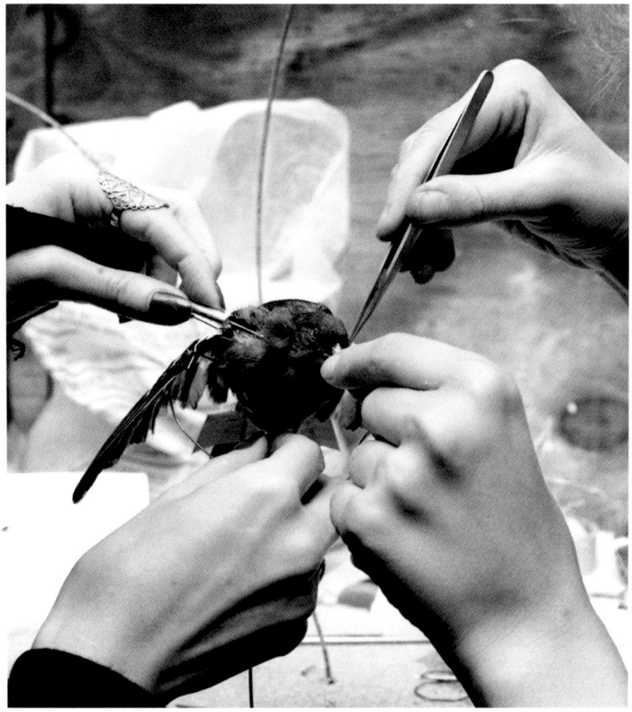

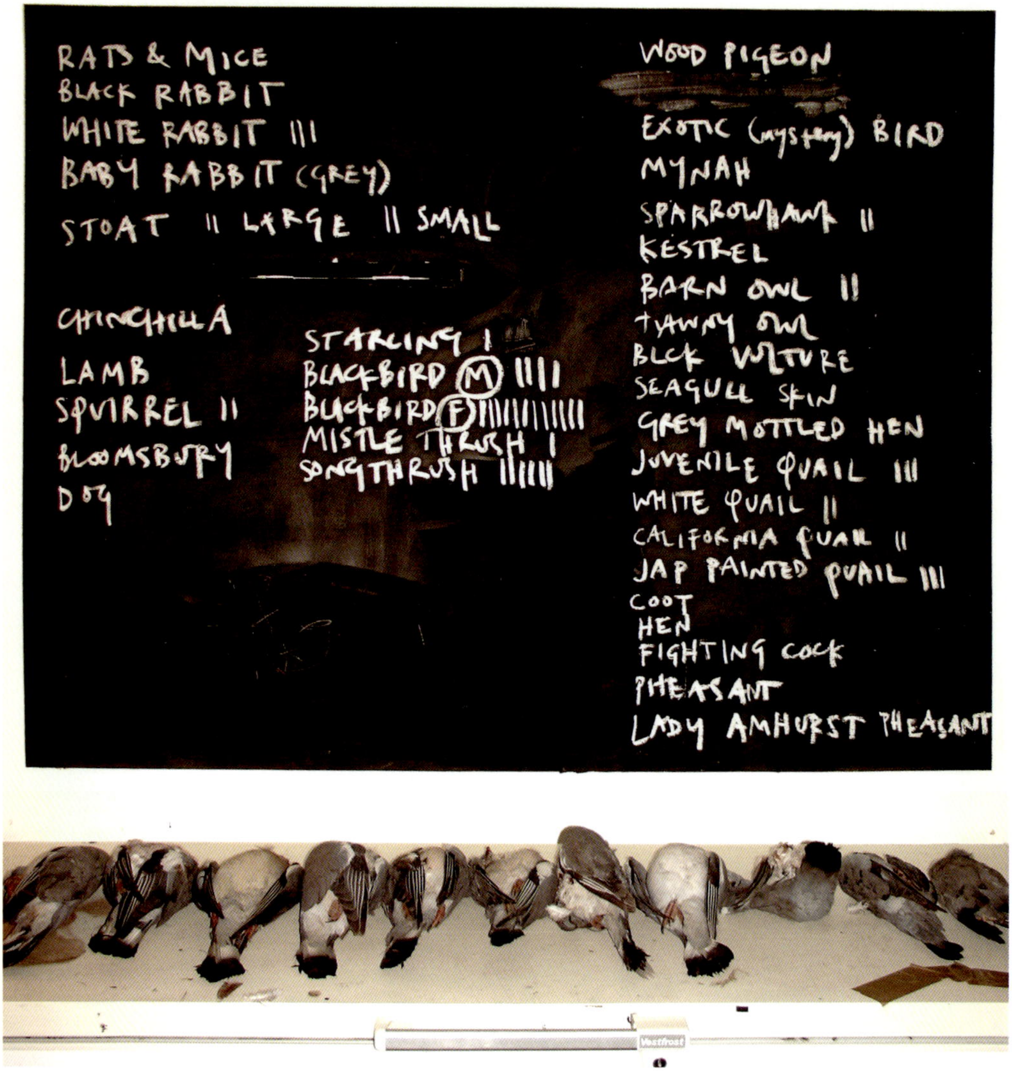
RATS & MICE
BLACK RABBIT
WHITE RABBIT III
BABY RABBIT (GREY)
STOAT II LARGE II SMALL
CHINCHILLA
LAMB
SQUIRREL II
BLOOMSBURY DOG
STARLING I
BLACKBIRD (M)
BLACKBIRD (F)
MISTLE THRUSH I
SONGTHRUSH
WOOD PIGEON
EXOTIC (mystery) BIRD
MYNAH
SPARROWHAWK II
KESTREL
BARN OWL II
TAWNY OWL
BLCK VULTURE
SEAGULL SKIN
GREY MOTTLED HEN
JUVENILE QUAIL III
WHITE QUAIL II
CALIFORNIA QUAIL II
JAP PAINTED QUAIL III
COOT
HEN
FIGHTING COCK
PHEASANT
LADY AMHURST PHEASANT
Vestfrost